# BODIES OF TOMORROW

Technology, Subjectivity, Science Fiction

Anxieties about embodiment and posthumanism have always found an outlet in the science fiction of the day. In *Bodies of Tomorrow*, Sherryl Vint argues for a new model of an ethical and embodied posthuman subject through close readings of the works of Gwyneth Jones, Octavia Butler, Iain M. Banks, William Gibson, and other science fiction authors. Vint's discussion is firmly contextualized by discussions of contemporary technoscience, specifically genetics and information technology, and the implications of this technology for the way we consider human subjectivity.

Engaging with theorists such as Michel Foucault, Judith Butler, Anne Balsamo, N. Katherine Hayles, and Douglas Kellner, *Bodies of Tomorrow* argues for the importance of challenging visions of humanity in the future that overlook our responsibility as embodied beings connected to a material world. If we are to understand the post-human subject, then we must acknowledge our embodied connection to the world around us and the value of our multiple subjective responses to it. Vint's study thus encourages a move from the common liberal humanist approach to posthuman theory toward what she calls 'embodied posthumanism.' This timely work of science fiction criticism will be of interest to cultural theorists, philosophers, and literary scholars alike, as well as anyone concerned with the ethics of posthumanism.

SHERRYL VINT is an assistant professor in the Department of English at St Francis Xavier University.

SHERRYL VINT

# Bodies of Tomorrow
## Technology, Subjectivity, Science Fiction

UNIVERSITY OF TORONTO PRESS
Toronto Buffalo London

Reprinted in paperback 2019

ISBN 978-0-8020-9052-2 (cloth)    ISBN 978-1-4875-2499-9 (paper)

---

**Library and Archives Canada Cataloguing in Publication**

Title: Bodies of tomorrow : technology, subjectivity, science fiction /
  Sherryl Vint.
Names: Vint, Sherryl, 1969– author.
Identifiers: Canadiana 20190126914 | ISBN 9781487524999 (softcover)
Subjects: LCSH: Science fiction, American – History and criticism. | LCSH:
  American fiction – 20th century – History and criticism. | LCSH: Science
  fiction, English – History and criticism. | LCSH: English fiction – 20th
  century – History and criticism.
Classification: LCC PN3433.8 .V58 2019 | DDC 808.83/8762–dc23

---

This book has been published with the help of a grant from the Canadian
Federation for the Humanities and Social Sciences, through the Aid to
Scholarly Publications Programme, using funds provided by the Social
Sciences and Humanities Research Council of Canada.

University of Toronto Press acknowledges the financial assistance to its
publishing program of the Canada Council for the Arts and the Ontario
Arts Council, an agency of the Government of Ontario.

Canada Council    Conseil des Arts
for the Arts      du Canada

ONTARIO ARTS COUNCIL
CONSEIL DES ARTS DE L'ONTARIO
an Ontario government agency
un organisme du gouvernement de l'Ontario

Funded by the    Financé par le
Government       gouvernement
of Canada        du Canada

Canadä

# Contents

# Acknowledgments

I would like to thank my PhD supervisor, Dr Doug Barbour, for his support and encouragement during my years working on my degree. Without him, I would not have been able to write this book. Thanks are also due to Heather Zwicker and Janice Williamson, members of my supervisory committee; their enthusiasm kept me at the project and their wisdom has improved it greatly. Jo-Ann Wallace is due thanks for the course on Feminist Theories of the Body that I took with her during my PhD program, a course that has inspired my thinking in this project and the years since.

In the years since the PhD, thanks are due to many people whom I've met in the field who have inspired and supported me. Most thanks is due to Veronica Hollinger, without whose encouragement I would never have gone to ICFA, a conference which has been rewarding to me professionally and personally. Veronica is a model of feminist scholarship of science fiction in whose footsteps I am proud to follow. Thanks also to Joe Sutliff Sanders, whose kindness and gift of Kelly Link's first book convinced me to continue to attend. Thanks also to Joan Gordon, another fine feminist scholar in the field, whose sage advice has improved this manuscript immensely. Finally thanks to UTP editors Siobhan McMenemy and Frances Mundy for seeing me through the long process of publication.

Most thanks, however, are due to Karys Van de Pitte. Without her friendship and support, this book would not have been possible. A sentence or two of acknowledgment cannot adequately convey the importance of hours of conversation over wine, that have helped me figure out many things, including some of the ideas in this book. Nonetheless, I offer them as the least I can do to let you know how much you are appreciated.

An earlier version of part of chapter 1 appeared as 'Double Identity: Interpellation in Gwyneth Jones's Aleution Trilogy' in *Science Fiction Studies* 28.2 (November 2001): 399–425. I thank the editors of *Science Fiction Studies* for permission to reprint this material.

BODIES OF TOMORROW

# Introduction: Problematic Selves and Unexpected Others

Science fiction is generally concerned with the interpenetration of boundaries of problematic selves and unexpected others.

Donna Haraway, 'The Promises of Monsters'

Greg Egan's short story 'Reasons to Be Cheerful' opens with the protagonist's announcement, 'In September 2004, not long after my twelfth birthday, I entered a state of almost constant happiness' (191). At first external reasons seem to explain this ebullient state, for his life is filled with 'food, shelter, safety, loving parents, encouragement, stimulation' (191); however, we soon learn that he has a brain tumour requiring aggressive surgical and chemical treatment. Although such treatment is able to benefit only two out of three patients, the protagonist feels 'no real panic, no real fear' (192). The reason for this, he tells us, is simple: a side-effect of the tumour is an elevated level of an endorphin, Leu-enkephalin, which binds to the same receptors as opiates; the protagonist is on a natural, permanent high. At great expense, his parents pursue a radical treatment that within three months eradicates the tumour and all chance of metastasis. Following treatment, the protagonist feels bad all the time, which at first appears to be a side-effect of adjusting to mere normal levels of endorphins, but which is ultimately revealed to be a side-effect of the treatment. The virus designed to kill the tumours also acquired the specifics of any cell designed as a receptor for Leu-enkephalin, meaning that 'every part of [his] brain able to feel pleasure was dying' (199).

By age thirty, the protagonist needs to take a daily cocktail of antidepressants that enable him to escape catatonia but still leave him too dysfunctional to do anything but be conscious. Yet another experimental treatment is proposed, this time injecting a tailored polymer foam into the damaged region of the brain, a polymer designed to function as a network of electrochemical switches. The foam and microprocessors scattered within it mimic and replace the lost neurons. Using this foam, his physicians install the composite data of neural connections for pleasure from about four thousand individuals. After surgery, all connections as experienced by any of the four thousand will be active, but experience will soon pare down those connections to reflect the 'real' preferences of the protagonist, using 'memories of formative experiences, memories of the things that used to give [him] pleasure, fragments of innate structures that survived the virus' to achieve 'a state that's compatible with everything else in [his] brain' (207). The narrowing process does not work, however. Upon awakening, the protagonist is returned to something akin to his twelve-year-old self's state of bliss, a condition in which all stimuli produce equal intoxication: 'All art was sublime to me, and all music. Every kind of food was delicious. Everyone I laid eyes on was a vision of perfection' (211).

The dilemma in the story is that this state of non-differentiated, all-inclusive bliss does not fit with the concept of individual identity. Although the protagonist has gone from a condition of abject misery to one of rapture, the fact that the neural connections producing the rapture are pre-wired based on many individuals rather than singularly produced via *this* individual's particular experiences renders the pleasure worthless in his eyes. Instead of feeling cured, he feels that 'all the joy I'd felt in the last ten days had been meaningless. I was just a dead husk, blowing around in other people's sunlight' (213). One more solution is attempted. The polymer neural net is modified with a set of controls that the protagonist is able to visualize. Rather than certain connections being reinforced and others fading through the vagaries of experience, the protagonist is able to choose for himself, 'consciously and deliberately, the things that make [him] happy' (213). Controls are installed and, after a brief training period, the protagonist re-enters society, able to decide his tastes for himself on a variety of axes (sexual, visual, acoustic, etc.) and also able to reverse any settings should he make a mistake or change his mind.

In the story's denouement, the protagonist finally decides he is ready to pursue a romantic relationship, and thus selects a woman – after a brief struggle about whether he should 'slaughter' (220) his heterosexual or homosexual self – for whom he will choose appropriate settings of attraction. The interpersonal relationship clearly demonstrates the anxieties about 'artificial' versus 'real' selves that trouble this story. He asks, 'How could I pretend that I felt anything real for Julia, when I could shift a few buttons in my head, anytime, and make those feelings vanish?' (224), yet also feels that 'it must be like this for everyone. People make a decision, half shaped by chance, to get to know someone' (224). When he first reveals his history to Julia, she seems to agree with the second reading, arguing, 'You chose me. I chose you. It could have been different for both of us. But it wasn't' (225). The next day, however, she breaks the relationship off, deciding that 'she wasn't prepared to carry on a relationship with 4,000 dead men' (226). In the end, the protagonist is left with his grief over the failed relationship – grief he decides not to turn off – and the struggle of continuing his life, having to face 'more starkly' than most that we all create selves out of this same legacy, 'half universal, half particular; half sharpened by relentless natural selection, half softened by the freedom of chance' (227).

This story raises a number of questions about identity, subjectivity, and embodiment that are central to the arguments of this book. At its core is the question of authenticity, of distinguishing 'true' from 'false' selves, of sorting out what is really 'me' from the programming of cultural influences on the one hand (in this story, the tastes of the four thousand individuals coded into the neural connections) and biological instincts on the other (raising the question of whether the child's endorphin-induced pleasures are any more authentic than the adult's hardware-induced ones). The story also points to the relationship between embodiment and subjectivity, raising the question of whether the same 'individual' is the protagonist in each section of the story. Certainly, his personality shifts radically as the chemicals and connections in his brain are shifted. As the story's conclusion suggests, the desire for an authentic self beneath the twin influences of biology and circumstance is a seemingly hopeless ideal yet not an ideal that we are quick to abandon. Although the story argues that 'normal' neural development and identity formation are a process that works similarly to the one undergone by the protagonist (albeit on an unconscious rather than conscious level), neither Julia within the story nor, I would suggest, most readers outside of it are convinced on a 'gut' level that both methods of development are the same thing.

Why are we not convinced? To put this another way, why are we attached to a concept of self that presumes an interior essence that is 'true' and immutable, or, within the terms of the story, why are we willing to accept the shaping and choosing processes as 'natural' and 'real' if they happen unconsciously yet inclined to see them as 'programmed' and 'meaningless' if the mechanisms of selection are brought to light? I suggest two reasons for this tendency. First, we are inclined to identify ourselves with 'voice' or self inside our heads, abstract essences that might be called souls in a religious context, but which persist also in non-religious concepts of self. This is the heritage of Cartesian dualism, a view of self that associates identity with the abstract realm alone. What is frightening about Egan's story is the notion that self might be merely the material, neural connections and neurochemicals that encode personality and memory.

Second, we value a concept of self as immutable and self-consistent, some essence that persists despite changes, including changes in our bodies. Lacan suggests that the process of acquiring an identity is a continual series of misrecognitions in which the ego is formed through what we identify with, what we desire. The fragility of this self, its ability to shift depending upon which objects of desire one chooses to identify with, is thus made visible by the story. It is no accident that the crisis point in the story is the love relationship, as the loved one is the ultimate substitute object to disguise the lack at the heart of identity. In order to retain a concept of ego or self as essential, unique, self-consistent, and autonomous, one needs to believe in the 'I' that does the choosing, that our choices of whom to fall in love with reflect some entity that precedes the moment of choosing. What Egan's story reveals is that this 'I' is formed by the act of choosing, that the 'self' who loves Julia is a self that can exist – or not – depending on adjustments in his desire for her. He is able to draw from a template of four thousand individuals' choices in forming the specific subject that he chooses to be, which stands in for the way cultural norms exist as templates for the rest of us while we make these choices, although 'normally' this happens on a level beneath conscious notice.

Egan's story thus highlights two aspects of identity formation that are vital to the arguments that I make in this book: subjectivity is as much material as it is abstract, about the body as well as about the mind, and subjectivity is shaped by cultural forces that produce the sense of an interior. In this book I will be looking at the relationship between subjectivity and embodiment that grounds our philosophical concept of selfhood. As the anxieties raised in Egan's story suggest, Western culture remains attached to a concept of self as disembodied, a concept of self that has important consequences for how we understand the relationship between humans and the rest of the material world, as has been noted by many theorists, including Latour, Haraway, Dery, and Graham. It is important to examine the consequences of this concept of self, for we are also living in a time when technology is able to radically alter the body through genetic therapies or cybernetic prostheses along the lines suggested by Egan's story (even if not yet to this degree). Technologies of body modification are developing and changing rapidly, provoking some to wonder whether 'by the end of the twenty-first century, there might well be no humans (as we recognise ourselves) left on the planet – but, paradoxically, nobody alive then will complain about that, any more than we now bewail the loss of Neanderthals' (Broderick, *The Spike* 15).

While the notion that twenty-first-century humans might be a new stage of evolution as different from us as we are from Neanderthals is an exaggerated view, the fact remains that technology is rapidly making the concept of the 'natural' human obsolete. We have now entered the realm of the posthuman, the debate over the identities and values of what will come after the human. On one level, this debate concerns the shape of the human future quite literally, referring to the embodied form of *homo sapiens* version 2 that might be produced through contemporary technologies of body modifications. Elaine Graham contends that 'what is at stake, supremely, in the debate about the implications of digital, genetic, cybernetic and biomedical technologies is precisely what (and who) will define authoritative notions of normative, exemplary, desirable humanity into the twenty-first century' (11). I would add that the outcome of such debates pivots greatly on the concepts of identity and embodiment that are dominant in the cultural milieu that surrounds the deployment of such technologies, and further that such values are significant not only for the effects they have on the human species but also for the relationships between humanity and the rest of the world that are implicit in them. My contention is that in thinking about the consequences of technologies of body modification, what is ultimately most important is the social milieu and philosophical assumptions which ground the way we deploy such technologies.

The purpose of this book is to explore the importance of embodiment to subjectivity and to look to science fiction as a privileged site that investigates some of the possibilities of changed embodiment for changing humanity. Although what is popularly compelling about the notion of the posthuman is the idea of a new physical way of being, what is more important and what underlies most science fiction engagements with the future of the human is a changed understanding of human identity. While many visions of the posthuman desire to transcend the limitations of the human body through technology or genetic redesign, I argue that it is important to return to a notion of embodied subjectivity in order to articulate the ethical implications of technologies of bodily modification. Technological visions of a post-embodied future are merely fantasies about transcending the material realm of social responsibility.

## Embodied Subjectivity

There is a tendency in some postmodern theory to speak of the body as an obsolete relic, no longer necessary in a world of virtual communication and technological augmentation. For example, Arthur and Marilouise Kroker ask, 'If, today, there can be such an intense fascination with the fate of the body, might this not be because the body no longer exists?' ('Theses' 20). They go on to argue that 'In technological society, the body has achieved a purely rhetorical existence' (21). Scott Bukatman argues that 'the body is not a requisite for the survival of the technocratic system' (*Terminal* 16) in which we now live, our embodied subjectivity having been replaced by a 'terminal' identity constructed on or in the screen. The idea of a natural body does seem to be rapidly eroding in a world where human DNA might be patented (Graham 119) and the list of cyborg implants which one might receive includes glass for parts of the ear; neuroprobes to monitor and medicate the brain; artifical glands, hearts, and pacemakers; penile and testes implants for cosmetic reasons; artificial joints; and implanted electrical stimulation to combat various autonomic function disorders (Gray, *Cyborg Citizen* 73). The human body also might be augmented or modified with various implants taken from animal sources, whether these be entire organs for transplant (still at an experimental stage) or various products of the new 'pharming' practice of using genetically modified animals to produce tissues such as skin, bone marrow, or nerve tissue for human treatment (Gray, *Cyborg Citizen* 122). These and other technologies of body modification have put into crisis the boundaries among human, animal, and machine. Changes to the body are one of the spaces where the posthuman may be literally made.

The ability to construct the body as passé is a position available only to those privileged to think of their (white, male, straight, non-working-class) bodies as the norm. This option does not exist for those who still need to rely on the work of their bodies to produce the means of survival, for those who lack access to technologies that can erase the effects of illness, and for those whose lives continue to be structured by racist, sexist, homophobic, and other body-based discourses of discrimination. The body remains relevant to critical work and 'real' life, both because 'real' people continue to suffer or prosper in their material bodies, and because the discourses that structure these material bodies continue to construct and constrain our possible selves. The material action of ideology on the body is not something that technology has erased; in fact, technology can be and has been used to enhance this action.[1]

Our failure to think of the body and its specificity as crucial to subjectivity is part of the legacy of Cartesian dualism. If we think of self as associated solely with mind, then technological changes to the body are not viewed as significant for human culture or human identity. Technophiles such as Hans Moravec have explored the possibility of making embodiment obsolete through the process of uploading human minds onto computer hardware, leaving behind the 'wetware' (in Rudy Rucker's terms) of the biological body. Another Egan story, 'Learning to Be Me,' imagines a world in which such a transfer is commonplace. In this story, every human is equipped with a jewel or artificial neural net that monitors and mimics the brain's processes. This jewel is trained to be 'you' until such time as you decide to make the switch to using the jewel instead of your organic brain. The theory is that 'so long as the jewel and the human brain shared the same sensory input, and so long as the teacher [a monitoring device] kept their thoughts in perfect step, there was only *one* person, *one* identity, *one* consciousness' (203). The story is the first-person narrative of someone who is very afraid of making this switch, fearing that having one's brain removed is akin to suicide and that, far from being immortal, he will simply be dead and replaced by 'some machine marching around, taking my place, pretending to be me' (206).

Despite his fear, the narrator allows himself to be convinced by the dominant logic of his culture that the switch to the jewel is transparent and decides to have the surgery. However, while he is awaiting his surgery date, he suddenly finds himself not in control of his own body, a mere spectator trapped in his head. The narrator realizes that, all this time, he has been the jewel, believing himself to be in control of the body simply because he was constantly being monitored and corrected by the teacher, erased each time one of his impulses did not match the choice made by the organic brain. Far from being the perfect imitation of the 'person' in the brain, the jewel is a separate individual, who comes to replace the other person whose identity is encoded in the organic brain. The synchronization is impossible to maintain 'naturally,' and ironically this is because of the very feature that makes the jewel desirable in the first place: unlike the organic brain, it never decays and so it inevitably operates differently from the organic brain.

There are a number of interesting conclusions about identity and embodiment to be drawn from this story. The most important, however, is that the idea of transcending human existence while still remaining 'the same' is clearly a fantasy. The body is what makes us mortal and weak, but it is this very vulnerability that should make us take care of ourselves, one another, and the planet we live on. The jewel decides at the end of the story that he has difficulty feeling sympathy for 'the man who spent the last week of his life helpless, terrified' because this man 'simply isn't *real*' (219). The jewel cannot sympathize with this man or feel him to be 'real' because the jewel has no way of 'knowing if his sense of himself, his deepest inner life, his experience of *being* was in any way comparable to [the jewel's] own' (220). Egan is clearly playing with the idea of the difficulties of intersubjective communication, the ways in which we are all trapped in our own skulls and unable to know if the Other has an inner life similar to our own, a problem that persists even if this Other is 'us' in the story. As in 'Reasons to Be Cheerful,' this story asks us to think about the construction of 'authentic' self. However, another implication of this story is that the model of subjectivity which is thus isolated and unconcerned about the other who is not 'real' is a model of subjectivity that refuses embodiment as an aspect of self. In the story, self is mind, and while this mind is different whether 'running' on the jewel or on the brain, the difference relates to the 'program' not the platform.

Thus, the concept of human identity that is implicit in the story is one in which the material world is resource and only the abstract mind has agency. The implications of this are important to our concept of human subjectivity in two ways. First, dualism is used to justify the exploitation of the material world and all those entities – which historically have included women and non-whites – who are deemed part of this 'natural' world of immanence rather than the cultural world of transcendent mind. Second, the notion of self as mind is linked to liberal humanist notions of identity, notions which rely on an abstract version of human sameness. Like the post-body versions of posthumanism, liberal human-ism posits a specific sort of embodied existence – which historically has

meant male, white, and propertied – as the 'essence' of all human identity. While there are positive aspects of this ideology in its attempts to
argue that all human beings are entitled to fundamental rights and
freedoms, what is often obscured by this ideology is the fact that certain
specificities are thus coded as 'outside' human identity, while others
that might be thought of as equally marked and specific are instead
taken to be transparent and universal. Returning the specificities of
embodied experience is one of the ways of resisting such erasures.

## Liberal Humanism, the Subject, and the Posthuman

One of my chief arguments in this book is that we need to pay greater
attention to the model of the subject that grounds how we think about
using technology to create the new version of the human, the posthuman. I have two central points of contention regarding the blindness of
some versions of posthumanism and the thinking about technology that
informs them. First, too often such models demonstrate the heritage of
Cartesian dualism and equate self with only the mind and ignore the relevance and specificity of embodiment. Second, the models of subjectivity
and society implicit in many of these models are still informed by the
assumptions of liberal humanism, particularly the unacknowledged ways
in which it has excluded certain subjects from its definitions of the
human. I argue that the persistence of some of these assumptions creates
lacunae in posthumanism and that unless we examine and counter these
assumptions, the versions of posthumanism we advocate will be guilty of
reproducing the same gaps which have troubled liberal humanism.

My critique of liberal humanist thought is limited to those aspects dis-
cernible in constructions of the posthumanist subject. As my primary
interest is in critiquing and offering an alternative to these models of
posthumanism, I will not provide an overview of liberal humanism but
instead isolate various elements of both liberal and humanist subject
construction – particularly their tendency toward false universalism,
abstraction from body, and distanced relation to nature – which persist
in negative ways in posthumanist thought. Both liberalism and humanism are multifaceted discourses, complexly interacting in liberal humanism, but rather than outlining this philosophical history, this book will
indicate how certain shared premises have been taken up in oversimpli-
fied and limited ways by some versions of posthumanism. Therefore,
throughout the volume, the term 'liberal humanism' does not connote
the entire field of thought but rather this particular cluster of elements

evident in much liberal, humanist, liberal humanist, and posthumanist thought.

My two key concerns with the liberal humanist tradition as it persists in posthumanism are the emphasis on universality and the emphasis on individuality. The emphasis on universality means that the vision of the subject embraced by humanism is taken to be timeless and based on some human 'essence' shared by all, which ignores the exclusions of women and non-Europeans (particularly non-whites) from the founding moments of both humanism and liberalism as theories of society. Tony Davies in his book *Humanism* points out that this notion of defining a shared essence of human nature is more an invention of nineteenth-century humanism projected back into the Renaissance, a view championed in particular by cultural critics such as Matthew Arnold (24). C.B. Macpherson's foundational book on liberalism argues that it is a philosophy of 'possessive individualism,' grounded in a bourgeois concept of society. Macpherson links the rise of liberal ideas to the political thought of Hobbes and Locke. While acknowledging the differences between these two thinkers, Macpherson draws attention to crucial assumptions they share: the vision of society as a series of market relations, a vision of the human as defined by propriety in his own person, and a definition of freedom as freedom from dependence upon others, freedom to benefit from the labour of one's own body and to own anything in nature that is shaped and changed by this labour.

As Macpherson points out, this model of society posits the individual as isolated in his self-ownership and as owing nothing to society for his person and his capacities (the gender-specific language is appropriate here), creating a society in which the only relations among men are relations determined by the market and exchange. This creates at the heart of liberal political theory a disparity between those who can own themselves and buy the labour of others and those who, because they lack other commodities through which to accumulate wealth, are forced to alienate a part of themselves through selling their labour. Liberalism's concept of freedom is troubled because it is an ideal produced in an era of uneven freedoms and thus its very concept of the 'human' subject (defined as free and entitled to certain rights) has always excluded some humans.

This ideal of the human as separate from nature and defined as human precisely through the subject's ability to shape and own nature is important to the humanist tradition as well as to the liberal one. This strict separation of humans from the rest of nature continues to trouble some versions of posthumanism which consequently continue to reproduce the blind spots that structured humanist discourses to the detriment of women, non-whites, and the working classes. Kate Soper points out that humanism, like liberalism, is founded upon a relationship of domination of the rest of the natural world, arguing that 'a profound confidence in our powers to come to know and thereby control our environment and destiny lies at the heart of every humanism; in this sense we must acknowledge a continuity of theme, however warped it may have become with the passage of time, between the Renaissance celebration of the freedom of humanity from any transcendental hierarchy or cosmic order, the Enlightenment faith in reason and its powers, and the "social engineering" advocated by our contemporary "scientific" humanists' (14–15).

This concept of nature as something separate from man and to be dominated by man combined with the concept of human society as based on market relations of exchange produces a vision of the individual as separate from the wider community. The concept of the subject as owning himself and owing nothing to society for this self or its capacities is evidence of a profound individualism that marks many version of the posthuman. This emphasis on individualism and isolation evacuates our model of society from any ethical sense of intersubjectivity and collectivity, which is also what I suggest is lacking from many models of the posthuman. Instead, we require a vision of the subject and the posthuman in which embodiment is central and self is seen as something that emerges from community rather than as something threatened in its autonomy by others. Thus, I will be arguing through my examination of embodiment in science fiction that the only viable posthumanism is one that goes back to the liberal humanist subject and starts anew, moving beyond the exclusions of the false universality of the humanist self and beyond the moral vacuity of the excessive individualism of the liberal self.

It is important to stress that in making this critique of liberal humanism, I am not ignoring or denying the many benefits that can be associated with humanism and liberalism, both in their moments of origin and as they have been taken up and modified by subsequent thinkers. Rather, I am trying to draw attention to an important structuring absence at the centre of each discourse, an absence that has to do with a distorting vision of universality and an overemphasis on individuality. My argument is that these exclusions and distortions persist in some of our models of the subject and particularly of the posthuman to a degree which has not been acknowledged, and that this means that posthumanism risks repeating some of the errors of the earlier discourses. Despite my concerns with some of the founding assumptions of these discourses, however, it is important to acknowledge the benefits that each brought to Western culture, the move toward secular society connected to humanism and the move toward expanded political franchise connected to liberalism. In what follows, I will be focusing on what is limiting about liberal humanism, but this should not be taken to be an argument entirely dismissing this system of thought.

Egan's short story 'Mister Volition' explodes the notion of the liberal humanist, autonomous subject who chooses, and also points to the ways that technologies are shaped by the cultural milieu in which they are deployed. In the story, the protagonist encounters a software product called Pandemonium that works on a patch, biofeedback technology. This technology 'can't show you anything that isn't there in your skull already' (93) but can make graphical displays of information that you might not otherwise consciously register. Patchware is a technology of self-shaping; it measures some state of the brain/body – stress, depression, concentration, arousal – and produces a graph of one's current state compared to 'some fixed template ... showing the result to aim for' (96). The protagonist is confused by the new patch at first because it shows only the patterns his own thinking produces, without providing a model to emulate.

The protagonist is also an existentialist, and spends a lot of his time thinking about his freedom to choose and the power and responsibility that come with this freedom. When he does choose a course of action, he is obsessed with understanding the agent of the choosing. He arrives at the conclusion that 'somewhere deep in my brain, there *must* be the "I": the fount of all action, the self who decides. Untouched by culture, upbringing, genes – the source of human freedom, utterly autonomous, responsible only to itself' (97). This 'I' is something the protagonist feels that he has always known existed, but he has been 'struggling all these years to make it clearer' (97). The technology of the patch enables the clarity he has been seeking. He concludes that what the Pandemonium patch does is show his thoughts in action, producing graphical displays that show all possible courses of action in a single moment, thought in action. The patch shows more than 'what I'd tell myself I'm thinking,' able to bring to consciousness 'all the rest: all the details too faint and fleeting to capture with mere introspection. Not the single, self-evident stream of consciousness – the sequence defined by the strongest pattern at any moment – but all the currents and eddies churning beneath' (100). The protagonist realizes that as he watches one pattern become dominant out of this pandemonium, he is watching the process of choosing. The patch can show him the self he has been searching for.

The protagonist discovers a pattern that surfaces 'whenever I thought of myself as a subject, an actor. Whenever I reflected on the power of the will' (102). This pattern he links to a test of absolute, existentialist freedom that he has set up for himself, the decision to kill someone just because he *can* choose. He names this pattern 'Mister Volition' and equates his self with this pattern 'controlling the body ... *making the decisions, pulling the strings*' (102, ellipsis and emphasis in original). It is important to note here that the protagonist equates self with this abstract pattern of thought, this concept produced in his brain, largely because he believes Mister Volition to be free, not subject, as he said before, to culture, upbringing, genes: to all the outside forces that shape the subject. It is significant that control of the body is one of the qualities he grants Mister Volition, mirroring the self-creation of Descartes. Self becomes that which is eternal, unchanging, autonomous, and in control, while all that is contingent, vulnerable, immanent, and subjected is labelled body and expelled from the self-concept. However, Egan's protagonist notices a problem with this software as he continues to investigate Mister Volition. He finds that 'I see most of the other patterns involved in the action flashing *first*, sending cascades converging on Mister Volition, making *it* fire – the very opposite of what I know is true. Mister Volition lights up the instant I feel myself choose ... so how can anything but mental static precede that pivotal moment?' (103).

When the protagonist confronts someone in the park and attempts to follow through on his experiment of achieving perfect freedom by shooting him, he finds that instead of 'the moment of lucidity, the moment of perfect understanding, the moment I step outside the flow of the world and take responsibility for myself,' he realizes 'Mister Volition is firing, but it's just one more pattern among thousands' (105). The insight he finally arrives at is that '*There is no first cause in here, no place where decisions can begin.* Just a vast machine of vanes and turbines, driven by the casual flow which passes through it – a machine built out of words made flesh, images made flesh, ideas made flesh' (105, emphasis in original). The protagonist is overwhelmed by this insight, but also ultimately made more human by it. He chooses not to shoot the man he

has confronted, and embraces a more complete view of himself as subject, not merely the will who chooses but 'the whole machine ... This
machine, and not another' (106).

There are three conclusions I want to draw from this story. First, the
story makes evident the relationship between concept of self and our
attitude toward and interactions with the rest of the world. Initially the
protagonist seeks a sense of self that is 'outside the flow of the world,'
and so long as he adopts this sense of self as abstract mind only, his
notion of responsibility and ethics is damaging and selfish. This sort of
self-concept is precisely what we must resist in the battle over posthuman identity. Second, the fuller sense of self that the protagonist
embraces, a sense of self that allows him to be part of the world rather
than its master and controller, is specifically an *embodied* sense of self.
The chief insight he has is that ideas are *made* flesh; they are not the
opposite of flesh. The necessary duality of self as embodied, specific,
'this machine and not another,' is also essential to the vision of the posthuman that I embrace. Finally, this short story and the others I have
been discussing make clear what science fiction can contribute to the
discussion of the posthuman. The story foregrounds connections between a technology and the cultural values that shape its design and use,
but also suggests that the technologies we use may shape and change us
in unanticipated ways. It is the ways that technology might change us –
both planned and unimagined – that make it essential that we think critically about the posthumanism we embrace in the twenty-first century.

**Body as Nexus**

Central to my arguments in this book is the contention that we need an
embodied notion of posthumanism if we are to return ethical responsibility and collectivity to our concept of self. The body occupies the lim-
inal space between self and not-self, between nature and culture,
between the inner 'authentic' person and social persona. Elizabeth
Grosz has defined the body as a Möbius strip, something which
'acknowledges both the psychical or interior dimension of subjectivity
and the surface corporeal exposures of the subject to social inscription and training; a model which resists, as much as possible, both dualism and monism' (*Volatile Bodies* 188). The strength of the Möbius strip
model is that it offers a way to conceive of the two aspects of the body
(interiority and surface) as always interacting yet not reducible to the
same thing, which allows analysis to address cultural inscription on both

the body and the subject, yet also looks for ways that the subject can resist such cultural marking and offer alternative possibilities.

The human body, like the human subject, is a product of both culture and nature. Both body and subject must maintain a sense of natural and stable boundaries by continually marking out the distance between what is self and what is not. The natural body is maintained through a number of boundary lines: that between male and female bodies, that between my body and the rest of the world, that between the natural body and artificial supplements to this body. These boundaries have always been unstable, and the recent abilities of technology to modify the body in radical ways make anxiety about these boundaries all the more apparent. Is someone who has had a sex-change operation a 'real' man or woman? Is an artificial limb part of my body or not? How much of my body can be removed or replaced while I still remain essentially 'me'? Does incorporation of a donated organ mean that someone else's body is in my body? The answers to these questions are not as straightforward as they might appear, but instead emerge from the interaction of the material body and the discourses through which we shape and make sense of this body.

The material and the discursive body are mutually productive: the material body is read by discourses, and the conclusions produced by these readings structure practices which influence the ways bodies come into being. In some instances, the role that ideology plays in producing the body is quite evident, as for example in the discourse of physiognomy which 'proves' the inferiority of non-white people or the discourse of hysteria which 'proves' the inferiority of women. However, the relationship is often more subtle and can more easily slip into being read as natural, such as the connection between discourses of female inferiority and the institutional practice of giving priority to men when distributing food, thereby producing smaller female bodies (which are therefore weaker, inferior, and need less food). Mind/body dualism continues to reassert itself in discourses, both those that challenge it – such as recent cognitive science which suggests that our use of metaphor in thought is radically structured by our bodies (Johnson and Lakoff) – and those that support it – such as legal principles which hold that an individual is less or not culpable for actions if his or her brain chemistry is somehow abnormal. How we feel about what is natural and what is not, whether the body is self or not-self, can have important consequences for social policy and personal political commitments in ways that are not easy to anticipate or predict. For example, in response to the search for a 'gay gene' many gays and lesbians are hopeful that scientific proof that homosexuality is 'natural' will reduce prejudice and persecution, while others fear that such a gene would be coded as diseased and eliminated through practices such as selective abortion (Terry).

In discussing the critical importance of the body, I wish to assess such interplay between nature and culture in its production, and to assess the extent to which the body is considered to be self or not-self. The solution to the first dilemma may seem obvious: clearly the body is 'natural,' something that we are born with, that 'comes' with certain abilities and features. However, this ideological obviousness soon breaks down under critical examination. In her essay 'Throwing Like a Girl,' Iris Young has demonstrated the ways in which socialization actively changes how a body is used and inhabited by an individual. As it turns out, girls do throw differently, but this difference is demonstrably a learned rather than a natural one. Similarly, Janet Stoppard has found that the symptoms of menopause are culturally specific and vary between North America and Japan. In *Volatile Bodies*, Elizabeth Grosz provides evidence that sufferers of multiple personality disorder may exhibit different bodily attributes from personality to personality, including such 'clearly organic' bodily functions as vision. The ideas that we have about what is natural or proper for our bodies influence what our bodies can and cannot do, and preconceived ideologies will determine what science will or will not find when it looks at them.

Ideology is the source of these various discourses that inform our ideas about our bodies and hence inform our experience of the lived body. In *Discipline and Punish* and *The History of Sexuality*, volume 1, Foucault outlines his theory of biopower regarding the ways in which social control of the body can be used to produce a specific type of subjectivity within that body. What we learn from Foucault is that the body is integrally linked to the discourses that make it intelligible. Biopower, with its classifications of normal or abnormal, valid or invalid, produces a field of hegemonic culturally intelligible bodies and produces bodies that fall outside of this field and hence cannot be 'seen.' The patch technology in Egan's story 'Mister Volition' is a perfect example of biopower, since the user is shaped according to the norm embedded in the patch, yet this shaping is done willingly, without the coercive use of force. The radical insight of biopower and Foucault's notion of the disciplinary culture that deploys it is that we willingly participate in our own subjugation; we must in order to become subjects at all. Judith Butler expands and puns upon this insight to classify bodies outside normative discourses as bodies which do not mat(t)er(ialize). This is not to argue that unintelligible bodies do not have a material existence, but instead to argue that such bodies can have no role in shaping the hegemonic ideology because they do not have a discursive existence. An example of this would be hermaphrodite births.[2] We have an ideology in which we conceive of the human race as divided into two separate and distinct sexes. When infants are born displaying the genital morphology of both sexes, the medical institution picks the sex that is dominant (that is, most 'true' to the nature of the infant) and surgically alters the 'deformity.' Hence, hermaphrodites are bodies that do not matter. An alternative set of discursive and institutional practices might conceive of sex as a continuum, and leave hermaphrodite bodies as they are born.

I am not trying to argue that either 'correcting' or accepting hermaphrodite morphology is a more 'natural' or appropriate response (for example, an ideology which 'accepts' hermaphrodite bodies could conceivably develop a very granular and discriminatory classification system for such bodies along the lines of race classification systems that distinguish between various percentages of 'coloured' blood). Instead, I am trying to suggest that the distinction between the material and the discursive body is a false one. Thus biopower is both the site of ideology's acting upon the body/subject and a potential site for resistance. Bodies which resist disciplining themselves to cultural norms challenge the field of the culturally intelligible. This is why the posthuman is a site of hope as well as concern for me, as technologies of body modification offer the chance to reshape bodies and thereby subjects and the social world we make; however, it is crucial that we understand the assumptions about the body and subjectivity that guide our technological choices in order to realize this positive potential.

## Politics and the Popular

SF is particularly suited to exploring the question of the posthuman because it is a discourse that allows us to concretely imagine bodies and selves otherwise, a discourse defined by its ability to estrange our commonplace perceptions of reality. The world itself must be imagined anew in SF and the conventions of the genre require the author to explore and explain the relationship between changes in the material world – which might include new technologies – and changes in the human subjects who inhabit this world.

Critics such as Russ, Lefanu, Barr, and Attebery have pointed out the utopian possibilities inherent in SF for rethinking gendered arrangements and assumptions. SF also has the power to make literal what is metaphor in other genres, a particular way of using language that is its distinctive feature according to Delany (*The Jewel-Hinged Jaw*). Science fiction has been posited as a genre that is particularly suited not only to exploring the world as it might have otherwise been but more specifically to exploring this difference *critically*. Freedman argues, 'The science-fictional world is not only one different in time or place from our own, but one whose chief interest is precisely the difference that such difference makes' (*Science Fiction* xvi).

Thus, science fiction is a space in which to explore the consequences of various versions of the posthuman within its imagined worlds. But SF is also important to my argument in the world outside the text as well. SF, like all cultural productions, forms a part of the world of available subject positions, of possible models for identification. This notion of subject formation explains the centrality of texts and representations in many arguments, including my own, focused on changing the social. If we can change the representations that are available for identification, we can change the subjects who are so produced. Thus, I am arguing for the importance of SF as a site of critical engagement with the discourse of the posthuman not only in terms of how SF texts might 'reflect' or 'illustrate' models of the posthuman theorized elsewhere, but also as a space in which models of possible future selves are put forward as possible sites for identification on the part of readers.

SF is part of the field of ideology and as such can work not only to comment on cultural politics of the current moment but also to intervene in and change this moment. Ideological representations are a normalizing strategy and, further, a normalizing strategy that is in the interests of the dominant class. By fostering normative identifications, the hegemonic ideology works to produce subjects who are suited to this ideology. In thinking of SF as having this power to potentially reshape the subjectivity of its readers, I am positioning SF as one of the discourses that 'calls' to subjects to interpellate them as concrete individuals as has been described by Althusser. Paul Smith argues that what is limited about Althusser's theory of interpellation is the absence of a space for agency and resistance, marked by a failure to theorize the effects of multiple, sometimes conflicting and contradictory, interpellations. In Smith's analysis, the subject is never entirely subjected to a given cultural formation, and space for resistance can be found in the fact that what interpellates the subject is changing and shifting. One of the places where such resistance might happen, where tensions between interpellations and cracks within the social context might become visible, is the activity of reading.

De Lauretis suggest that cultural practices can have an even greater revolutionary potential through what she calls reverse discourse, which is 'the process by which a representation in the external world is subjectively assumed, reworked through fantasy, in the internal world and then returned to the external world resignified, rearticulated discursively and/or performatively in the subject's self-representation – in speech, gesture, costume, body, stance and so forth' (*The Practice of Love* 308). SF texts are a place where reverse discourses of the dominant notion of what it means to be human are continually being performed. As the possibilities for material subjectivity interact with the possible subject positions presented within the culture, part of the work for social change is challenging and changing the representations that circulate within the currently dominant construction of the social.

As we move into the twenty-first century, a number of fantasies of the posthuman identity are at play in fictional and technoscientific cultural representations. Which of these posthumans will form part of the new hegemonic ideology depends on social and cultural conditions for the use of technologies of bodily modification, a choice we are in the process of making. It is clear that social change will take place in the twenty-first century as we struggle to come to terms with the implications of AIDS, the Human Genome Mapping Project, and new reproductive technologies. The goal of this project is to look at the ways that certain science fiction texts have mapped out where change could take place, for better or for worse, through an examination of how they represent the relationship between the body, the self, and the social.

The new selves SF might help us imagine are both the problematic selves and the unexpected others of this chapter's epigraph: they remind us of the fragility of our boundary-making work and that the Other always *is* an aspect of self made problematic. As we explore the posthuman, our future bodies and future selves, I am aware of the need to be cautious. Technology is neither emancipatory nor repressive in and of itself and can be used to signify new forms of exclusion as well as new spaces of freedom. The creation of new posthuman bodies with technology involves social and moral as well as technical choices. It is my hope that this reading of these science fictional texts provides both opportunities for exploration and warnings for routes not to be taken, as we continue to struggle with the implications of new technologies of the body in this current cultural moment.

**Future Bodies**

Although set in the future or elsewhere, science fiction is commonly understood to be about the moment contemporary to its production, the anxieties and anticipations that form that moment. As I am interested in interrogating current anxieties about embodiment and posthumanism, the texts I examine have all been published within the past twenty years. Similarly, since the purpose of this study is to explore ways in which we can rethink the concept of human identity and embodiment with the aid of science fiction examples, I have selected texts and authors based on the philosophical problems highlighted by particular texts rather than with reference to national boundaries or other ways in which these texts might be put into literary rubrics. Finally, I have chosen these texts as representative examples of how certain themes are treated within SF in order to elucidate a range of concerns about reading and writing technologies, and posthuman futures via genetic or information technology; clearly other texts or additional texts might have been chosen.

Chapter 1, 'Gwyneth Jones: The World of the Body and the Body of the World,' explores issues about the body, subjectivity, and ethics through a reading of Gwyneth Jones's trilogy of novels, *White Queen* (1991), *North Wind* (1994), and *Phoenix Café* (1998). The chapter provides the groundwork for my argument that textual representations are an important source of identifications in the construction of self and hence may also be an important site of resistance to this process of assuming a culturally intelligible identity. I read the alien characters' cultural practice of 'becoming yourself' in terms of Althusser's theory of interpellation and Judith Butler's work on performative subjectivity. Identity is a term that connotes both sameness and difference, and I argue that the novels' engagement with these multiple significations of the term suggests a model of subjectivity that is not rooted in repudiating its others.

Chapter 2, 'Octavia Butler: Be(com)ing Human,' explores the posthuman bodies that are promised by the new genetic technologies. Within the context of critiques about socially discriminatory uses of genetic information, I provide a reading of Octavia Butler's *Xenogenesis* trilogy (recently renamed *Lilith's Brood*): *Dawn* (1987), *Adulthood Rites*

(1988), and *Imago* (1989). These novels are not rooted in a 'hard science' of genetic technology; rather, they are explicitly concerned with the way human culture has historically read and judged people through their bodies. Butler's nuanced exploration of the fear humans feel about having their genetic identity altered confronts us with the dangerous implications of gene therapy and its ability to write social readings of 'normal' bodies and subjects onto the body at the genetic level. The trilogy reminds us directly of the material consequences of the intersection of bodies, ideology, and technology.

Chapter 3, 'Iain M. Banks: The Culture-al Body,' explores a model of posthumanism that I deem too disconnected from embodiment particularly because of its unacknowledged allegiance to liberal humanist notions of human identity. Banks's novels *Consider Phlebas* (1987), *Use of Weapons* (1988), and *The Player of Games* (1990) explore a kind of posthumanity in which the body has become an infinitely malleable accessory. While on the surface this future appears a place of infinite variety – one can adopt any among a spectrum of genders or racial features – a closer look shows that this variety conceals an underlying essential similarity. Both Banks's Culture and liberal humanism share a naïveté about achieving equality by erasing difference via positing the body and its experiences as irrelevant to subjectivity. Unacknowledged disparity, not equality, is created when we erase bodily specificity from the social order.

Chapter 4, 'Cyberpunk: Return of the Repressed Body,' turns to cyberpunk fiction, renowned for its rejection of the body as mere meat and its celebration of the freedoms of disembodied subjectivity in cyberspace. I link cyberpunk's fear of the body to Cartesian mind/body dualism and then this dualism to the universal man of liberal humanism. The cultural context of cyberpunk's emergence, the 1980s, includes the threat of human obsolescence in an increasingly automated workplace, the increased globalization of capitalism, and an increasing gap between the rich and the poor. This context offers an explanation as to why escape from the body might appeal, and I argue that Gibson's *Neuromancer* (1984) shows us just this fact rather than celebrates the cybercowboy identity, as is often assumed. I contrast this fictional version of life in cyberspace with Ellen Ullman's *Close to the Machine* (1997), a software designer's autobiographical reflection on her 'cyber' life, which shows a similar ambivalence about the material world over the cyber one, and a similar awareness of the material reasons that lie behind this ambivalence. Finally, I conclude the chapter with Pat Cadigan's *Synners*

(1991), a novel that is often considered to be the 'feminist' response to cyberpunk. Cadigan makes embodied reality and subjectivity central to the ethical dimension of the text: her concern is with responsible use of technology and its accompanying cultural changes. I juxtapose Cadigan's argument for the responsible use of technology with Marxist critiques of the concept of humanware in current industrial projects. Humanware is the notion that human workers can be treated as just another component in industrial systems, a marketplace logic that insists upon subordinating the human 'components' to the logic and pace of the machine.

Chapter 5, 'Raphael Carter: The Fall into Meat,' provides the final look at cyberpunk's repression of the body in *The Fortunate Fall* (1996). This novel is an ironic response to the cyberpunk formula, using its expected tropes but undermining their efficacy and thereby denying cyberpunk's claims to revolutionary potential. In Carter's novel, the material world remains the important site of contest, and characters forget this at their peril. Carter refuses to let the reader cling to the romantic notion that a heroic, single cyber-cowboy can defeat large institutional forces of oppression through ingenuity and skill. Those who live only on the Net are shown to be less instead of more than human. In these and other ways, Carter's novel shows that the body is necessary to ethical interactions with the world and that a post-body posthumanism will necessarily be oppressive. The novel further suggests that the forces of repression will ultimately win so long as they remain capable of controlling representation.

In chapter 6, 'Jack Womack and Neal Stephenson: The World and the Text and the World in the Text,' I return to the cultural politics of representation. Here I argue that representations matter to subject formation and to efforts to connect with and change the cultural politics of the material world. This chapter explores reading and writing as technologies of self and world transformation in Neal Stephenson's *The Diamond Age* (1995) and Jack Womack's *Random Acts of Senseless Violence* (1993). In Stephenson's novel, the electronic *Primer* allows the protagonist, Nell, to rewrite herself, but she also learns that textual representations are not sufficient for social change. It is only by critically reflecting on the gaps between the *Primer's* representations and her own experience that Nell learns to extend her new self beyond the text and thus remake her world. Womack's novel, written in the form of its protagonist's, Lola's, diary, narrates her attempts to cope with the changes in her life as economic pressures force her middle-class family into a life of poverty. Lola uses the diary to try to remain attached to her old material life; however, the self she writes in the diary is ultimately incapable of coping with the new, viscerally violent world into which she is thrust. Those around her refuse to accept her reading of herself and instead insist upon imposing their own social reading of her worth and prospects. In the end, Lola abandons both the diary and the hope of recovering her old life.

By comparing these two novels, I argue that we can see the possibilities and limitations of textual interventions in the social order. Lola and Nell achieve very different results in their attempts to reshape themselves through texts. What these novels reveal is the importance of a community of readers who accept and act on the text's new representations. Without community and the material world, ethical posthumanity is not possible because the move toward posthuman identity will be grounded on disconnection from the rest of the world and our ethical responsibility to it. Only by articulating our posthumanism within a social network can we find a non-solipsistic way to move beyond our current concept of what it means to be human.

The book's conclusion, 'Toward an Ethical Posthumanism,' argues for the importance of contesting competing visions of future humans and specifically that it is essential for embodiment to figure in our understanding of the posthuman subject. I begin with a discussion of Vernor Vinge's idea of the singularity, defined in his essay 'What Is the Singularity?' Vinge argues that we are technologically on the cusp of an event whose precise shape is as yet unknown but whose emergence will mark the end of the human era and the beginning of the next. Once this singularity occurs, he suggests, the entire world will be different and this posthuman future might simply be incomprehensible to humans who have become irrelevant to it, superseded by the 'next' of human evolution. As the body of my book shows, there are considerable ethical dangers in some of the models for the posthuman currently being considered in both fiction and technoscience practice. However, I suggest that if we model our ideals of the posthuman on moving beyond liberal humanism, then there exists the more positive model of an embodied posthuman subjectivity. This model, I suggest, can see posthuman subjectivity in terms of its possibilities for multiple forms of embodiment, an embodiment that is continually changing and open to new ways of engaging with the world as we experience it from multiple subject positions.

Elaine Graham reminds us that we need not fear the posthuman. She points out that 'The "end of the human" need not necessarily entail a choice between "impersonal deterministic technologized posthumanism" and "organic unmediated autonomous 'natural' subjectivity," but may involve modes of post/humanity in which tools and environments are vehicles of, rather than impediments to, the formation of embodied identity' (199). My engagement with posthumanism throughout this book is similarly concerned with exploring some of the ways that we might be posthuman *and* embodied, and why embracing such a configuration might matter. Embodied posthumanism has the power to expand our capacity for responsibility and our connections with others. It is inevitable that technoscience will continue to enact changes on the current state of human embodied existence. Thus it is imperative that we develop an ethically responsible model of embodied posthuman subjectivity to ensure that such bodily modifications make us more than – rather than less than – human.

# 1  Gwyneth Jones: The World of the Body and the Body of the World

The awakened and knowing say: body am I entirely, and nothing else; and soul is only a word for something about the body.

Friedrich Nietzsche, *Thus Spoke Zarathustra*

Gwyneth Jones's *Aleutian* trilogy provides an exemplary set of texts exploring the intersections of the subject, the body, the text, and the social. In this trilogy, Jones provides a new model of the body, a model premised on the deconstruction of the boundaries between human and alien. This new image of the body becomes a ground for a new kind of ethics in Jones's work. The body is important for understanding this theme because it functions as both a tool for articulating self and as a conduit through which cultural meanings shape the body/subject. Through a new understanding of the body, Jones's work suggests, we are able to create a new version of the social, a more ethical social world which does not insist upon forming self through repudiation of the Other.

The trilogy – *White Queen* (1991), *North Wind* (1994), and *Phoenix Café* (1998) – describes a three-hundred-year period that spans from the first contact with an alien culture (the Aleutians),[1] who arrive on earth in 2038, until their departure from our planet. *White Queen* narrates the period of first contact and deals primarily with the misconceptions and misunderstandings that occur because both the humans and the Aleutians insist upon reading the other through the standards of their own culture. The humans assume that the Aleutians are a superior race who has arrived by faster-than-light (FTL) travel, a misperception that is fuelled by the Aleutian idea of reincarnation. The humans believe that the Aleutian individuals are the same *physical* individuals who departed from their distant planet, while the Aleutians at first assume that the humans share their understanding that reincarnated subjectivities are the same *personality* across the generations. The Aleutians view the humans as potential trading partners and see their relationship in terms of opportunity for profit rather than in terms of the human perception of interplanetary intrigue and domination. The main human characters in this novel – Braemar Wilson and Johnny Guglioni – work to uncover what the aliens are 'really' up to, while the central Aleutian character – Clavel – believes that he falls in love with Johnny. The cultural misunderstandings culminate in the 'rape' of Johnny by Clavel.[2]

*North Wind* is set approximately a hundred years after first contact. This novel recounts the race between a number of parties – Aleutian and human – to recover the technology of FTL travel that was discovered and hidden by a human character – Peenemünde Buonarotti – in the first novel. Peenemünde had hidden her discovery, as she believed that only Aleutians would be able to use it[3] and she did not want to give the aliens another advantage over humanity. The central character of this novel is Bella, an Aleutian who is confused about her identity. It is ultimately revealed that Bella is an Aleutian person who has been created through genetic engineering from Johnny's human tissue. The Aleutian character Clavel – returned in another incarnation during this time – had hoped that the misunderstanding of the rape could be healed in another generation through this creation of a new Johnny.

The final novel, *Phoenix Café*, is set another two hundred years into the future. Over this time period, relations between the Aleutians and the humans have become increasingly strained. Although it was not their original intent, the Aleutians find themselves in the position of colonizer; as the result of their superior technology, they have usurped many of earth's political powers. This novel narrates some of the long-term consequences created by the initial misunderstandings upon which the relationship between humans and Aleutians was based. There are three main narrative threads in this novel: the attempt of the humans to use Aleutian organic weapons technology in their gender war; the discovery of FTL travel that humans may use by a human character, Helen; and a final attempt to overcome the damage of the rape incident through the character of Catherine. Catherine is the reincarnation of Clavel in a human body.

Identity is a central theme of these three novels. Identity is a term that can mean either that which uniquely identifies or determines the 'essence' of the thing named, or the condition of being indistinguishable from or identical to something else. Both of these denotations are explored by the trilogy as Jones plays with the contradictory meanings of the term. The characters struggle to negotiate their sense of self, that is, those unique characteristics which make 'me' who I 'am.' The humans struggle with the need to reaffirm the boundaries of what is human, as the existence of the, presumably superior, aliens challenges their sense of their place in the universe. Finally, Jones uses her trilogy to demonstrate the damage that results from the need to construct identity on the ground of repudiated Others, suggesting ultimately that – after all – there is identity in the sense of sameness, between human and Aleutian, and between people and the rest of the world.

The relationship between ideology and identity formation is a familiar one to those conversant in critical theory. As Jones argues in her introduction to *Deconstructing the Starships*, the concern of critical theory with 'plurality of meaning, fluidity and process: an understanding of language as contingent, unfixed; the product and definition of a particular social formation' draws on concepts which are 'the familiar tools and usage of science fiction' (3). The role of ideology or cultural representations in the process of subject formation is central to contemporary theories of the subject. These theories address three central issues: the tension between nature, or innate traits, and nurture, or ideological influences, in identity formation; the mechanism by which ideology or culture can be understood to work on the subject; and the role of repudiation or refusal of certain identities in the construction of subjectivity. Jones's novels address each of these three aspects of identity formation through a first-contact narrative exploring thematic issues similar to those raised by debates about subject formation. How does ideology come to confer an identity upon us? What are the consequences of this relationship between ideology and identity for our relations with other subjects? And what room remains for resisting or transforming hegemonic ideology if it is the very source of our identity?

Althusser has argued that '[i]deology has the function (which defines it) of "constituting" concrete individuals as subjects' (45). He explains that ideology is both that which allows us to recognize ourselves and that which allows others to recognize us. Ideology, then, is central to subject formation and to the continuation of social orders. Ideology continually (re)produces the subjects which are suited to its current social institutions, as the calls of those institutions are answered by individuals who view this call as addressed to 'really me,' that is, the 'real me.' Ideology thus produces both subjects and the social order into which those subjects fit. Biases and discriminatory categories of the social order are rendered 'natural' by the workings of ideology and the process of interpellation. An important part of Althusser's theory is his conception of what I will call the sincerity of this recognition. This is not the false imposition of an identity by an external force, but is rather the acknowledgment by the willing subject that he has finally found his 'true' place in the social order, one that 'fits' with the interior essence that he believes to be himself.

Judith Butler's work on the formation of gender identity expands this Althusserian model in useful ways. Most importantly, as Butler is at pains to remind us in both *Bodies That Matter* and *The Psychic Life of Power,* the subject *emerges* in this moment of recognition, the subject is formed by this moment. It is only the illusion of grammar, Butler argues, that gives us the impression that the subject precedes its answer to the call. The 'something' that turns in recognition is only retroactively installed in our grammatical construction of the event. Butler analyses the importance of thus accepting a place in the domain of culturally intelligible subject positions – those identities which are described and acknowledged by the hegemonic ideology – as something that constrains the subject's very ability to exist at all. What is sometimes overlooked in discussions of subject formation is the extent to which the subject is formed by both identifications and repudiations. Just as the subject comes into being by answering the specific call that was made 'to him,' the subject also refuses to acknowledge those calls that are made to someone else.[4] As subjects are able to enjoy intelligibility and hence social status only to the extent that they *are* recognized by the culture in which they emerge, it becomes very important for subjects to repudiate identifications or calls that would serve to jeopardize the subject's cultural intelligibility.

As Butler acknowledges but Althusser overlooks, the subject is formed by multiple and competing calls related to gender, ethnic identity or race, sexuality, and class. Anxiety emerges when the stability of social categories is challenged, and this instability is generally revealed by those subjects who do not easily 'fit' into one category or another. It is critical for the subject to disavow any continuity between itself and its constructed others, since a threat to this boundary is a threat to the subject's very conception of itself. The role of repudiation in subject formation explains in part the vehemence which underlies discourses of sexism, racism, and homophobia. The threatening truth that the other could be us menaces our sense of security in our own identity. It is therefore important to foster identifications that resist these disavowals. These repudiations inform the discourses that socially divide people, and we can change these discourses only through changing the identifications and disavowals.

Althusser's model of interpellation is very static. Butler believes that the model of interpellation can be restructured to offer the possibility of resistance and change because of the material nature of ideology which is embodied in the institutions *and practices* that *continually reproduce it.* This reproduction through repetition is key to Butler's understanding of how ideology works to form the self. The norms or subject positions that come into existence in hegemonic ideology can retain their materiality only to the extent that they remain embodied in the practices of individuals and institutions. Butler argues that '[t]he social categorizations that establish the vulnerability of the subject to language are themselves vulnerable to both psychic and historical change' (*Psychic Life* 21). The signifiers that describe intelligible identity in a given time and place are 'capable of being interpreted in a number of divergent and conflictual ways' (96). Butler argues that interpellation can never successfully encompass all that exists before the subject emerges from the process of interpellation; there is always a remainder, that which cannot be expressed within the current limits of the culturally intelligible and which therefore does not appear in representation.[5] This excess, this unnameable, is what returns to subvert the success of interpellation.

Jones's trilogy contributes to this debate about the relative strength of the cultural forces of ideology and the potential space for human agency. First, it provides a representation of interpellation as alien technology, a representation that can help the reader see the 'obviousnesses' as socially constructed and thus recognize the call as a call that may be either answered or resisted. Second, the novels perform their own ideological call for a new kind of subjectivity. Through the shifting meaning of the concept of identity in these novels, Jones describes a human subjectivity that recognizes its connections to other living beings in the world, a more ethical subjectivity than the traditional Western concept of the autonomous and self-contained individual. The creation of this new type of subjectivity was one of Jones's central aims when she set out to write these novels; she wished to create 'complex and interesting people who managed to have lives fully as strange, distressing, satisfying, absorbing, productive as ours, without having any access to that central "us and themness" of human life' ('Aliens' 111).

One of the reasons that this set of novels is so useful for exploring the ways in which identity is connected to culture is that Jones has chosen to narrate her novels from the points of view of both the humans and the aliens.[6] Thus *White Queen* provides us with only a provisional insight into the nature of the alien race and their culture. The reader, with the human characters, searches for clues that will decode the 'meaning' behind their acts and statements. At the same time, portions of the narrative that present the aliens' point of view perform the same interpretative work to decode human behaviour. This style of narrative makes apparent the difference in strategies used by the aliens and the humans to attempt to read the other.

The humans read the Aleutians through an assumption of difference while the Aleutians read the humans with an assumption of similarity. Throughout the novels, many misunderstandings follow from this initial distinction: the humans assume that the Aleutians are superbeings with powers of telepathy and FTL travel, while the Aleutians assume that they are involved in trade negotiations with 'the locals,' simply business as usual. The Aleutian confusion is revealed in the following exchange between two Aleutian characters:

> <What I can't understand … Is how they came to be expecting us.>
> <It wasn't us they were expecting. It was some other, important people.>
> (*White Queen* 93)[7]

This aspect of cultural misunderstanding reveals the degree to which we perceive the other through our own cultural preconceptions. This crucial distinction – the humans expect difference while the Aleutians expect identity – structures the relationship between the two species.

The novels emphasize the human attempt to gather data on the Aleutians in order to understand their difference. In *White Queen,* this desire for information emerges from a perceived need to understand whether the Aleutians pose a threat to humanity. *Phoenix Café* roots this search in a desire to reassert what is fundamentally different and unique about being human. The humans enter the relationship based on the assumption that the aliens are superior: 'It was a truism that the aliens who landed, whoever they were, *had to be superior.* Or else we'd be visiting them' (*White Queen* 71). The humans observe that the aliens communicate without verbalizing and interpret this as a sign of alien telepathy. Gradually, the novels reveal this ability as something called the Common Tongue, which is a combination of body language, cultural conditioning, and biochemical feedback.

An understanding of the Common Tongue is essential to understanding how Aleutian culture works and to understanding the differences that exist between humans and Aleutians. Jones forces the reader to struggle for this understanding, allowing evidence to emerge only gradually throughout the trilogy. Initially, those characters who reject the notion of the Common Tongue as telepathy believe that it is a sophisticated type of body language. Braemar describes the phenomenon in this way: 'Babies don't learn to speak in order to communicate. They get on perfectly happily without words as long as they're with people who know them ... Gesture, body language: when you know someone well, an educated guess. That's what the Aleutians have' (*White Queen* 179). Through studying the alien body, the humans learn that the aliens' DNA performs differently from human DNA. In humans, most DNA is 'junk' and DNA fingerprinting can be used to establish the unique identity of an individual. For Aleutians, each nucleus includes the potential for expressing three to five million unique individuals. Aleutian individuals are produced by 'a chemical event – analogous to our "moment of conception" – forever deciding which of the strings is expressed' (*White Queen* 168).

Rather than relying on biology to determine unique identity, aliens turn instead to culture. This is the process of 'learning to be oneself' that forms the final component of the Common Tongue. As Braemar's comments above indicate, the ability to communicate without language exists for infants provided that they are among people they know well and who know them well. In Aleutian society, we discover, everyone knows everyone else well, through intense study of the 'character records' of their own and other people's lives. Aleutians do not believe in 'permanent death'; they believe that the unique genetic expression for each individual will return in the next generation. The reincarnated person will learn to become his self[8] through studying the records of his previous lives, awaiting the 'chemical event' of recognition that allows the individual to know which of the people represented in the records he 'is.' Once this moment of recognition occurs, the individual models his life on the example provided by the records, and makes his own records for the edification of future reincarnations of his 'self.'

The Common Tongue is thus positioned at the intersection of nature and culture. It is a reading of the body, both its gestures and its biochemical composition (through the Aleutian practice of ingesting one another's semi-sentient mobile cells called wanderers); but it is a reading of the body that can only be made by someone who is formed by the ideology of the culture. Aleutian identity is a good example of the kind of embodied subjectivity that Grosz calls for; it encompasses both sides of the Möbius strip. Aleutians experience the biological moment of recognizing themselves during the cultural study of character records.

The Aleutian process of learning to be oneself can be thought of as a technologized version of interpellation. The subject is formed at this moment of answering the call, of recognizing that the call was 'really' for him; for Jones's Aleutians, this process takes the form of a religious ritual that includes both a cultural and a biochemical component. When an Aleutian recognizes himself in the cultural record, this recognition takes the forms of both psychically recognizing the representation as something that corresponds to the 'real' me, and physically experiencing the biochemical secretions of the record as a call to one's unique biochemical identity. Aleutian identity is thus formed by both culture (the record study) and nature (the 'chemical event' that triggers recognition of the self in the records), challenging the human separation of mind from body. Once this moment of recognition has occurred, the Aleutian becomes a disciplined body, working to model his self on the records of this 'self' left by previous incarnations. However, like Althusser's subject who recognizes that it 'was really me' who was called, the Aleutians do not perceive this practice of learning to be one's self as an imposition of outside ideology; rather, it is the fulfilment of their inner, true identity.

This Aleutian practice occupies a space of tension in the novels between the notion of identity as that which uniquely characterizes and the notion of identity as sameness. For many human characters, the Common Tongue in both its biological and social components is that which distinguishes humans from Aleutians: they are serially immortal, we die once; they contain the whole brood's DNA, we have a unique fingerprint; they share biochemical unity with their brood, we are separate individuals. Throughout the novels, however, humans and Aleutians are able to communicate through the Common Tongue, albeit a somewhat muted form because human Common Tongue lacks the biochemical exchange. In fact, Common Tongue is portrayed as a 'natural' part of human communication that humans are simply not trained to read. By the time period of the third novel, *Phoenix Café*, many humans have learned to repress their speech in Common Tongue in order to maintain their sense of privacy and Aleutians have learned to ignore many humans statements in the Common Tongue given their inadvertent rudeness. As well, some humans respond to the arrival of the aliens by imitating them and adopting Aleutian practices such as looking for and learning to be oneself in visual records.

Humans who imitate Aleutian culture are called halfcastes. These halfcastes, the name invoking both racist and classist discourses, are a cultural rather than biological miscegenation. As they embody aspects of both human and alien they are often the targets of hatred and violence. Jones uses the halfcastes to illuminate the ways in which border-crossing identities raise anxiety in those whose selves are predicated upon 'us and them' distinctions. If human identity – figured in terms of that which makes us uniquely ourselves – is constructed out of both identifications and repudiations, then anyone or anything that occupies the margin of self and other threatens the (constructed) existence of the self. The very existence of halfcastes suggests that humans and Aleutians are not as different and separate as we might first imagine. Jones plays with the idea of identity shifting from 'marking uniqueness' to 'connoting sameness' through the halfcastes. Through this play, she undermines the human characters' construction of themselves as distinct from the aliens and suggests a new way to conceive of humanness through what we share with our others.

In her essay 'Aliens in the Fourth Dimension,' Jones describes her 'laboratory testing' of the Aleutian species she has created from her premises:

> They were an absolutely, originally different evolution of life. But they were *the same* because life, wherever it arises in our middle dimensions, must be subject to the same constraints, and the more we learn about our development the more we see that the most universal pressures – time and gravity, quantum mechanics; the nature of certain chemical bonds – drive through biological complexity on every fractal scale, from the design of an opposable thumb to the link between the chemistry of emotion and a set of facial muscles. This sameness, subject to cultural variation but always reasserting itself, was shown chiefly in their ability to understand us. (116)

In this passage, Jones demonstrates the strength of her conviction that ideologies that separate and classify us are misleading and dangerous. This similarity between humans and aliens allows us to understand Jones's depiction of the process of learning to be oneself as a representation of the process of acquiring a culturally intelligible identity.

We must remember that interpellation constructs the subject in accordance with the dominant ideology. The records that Aleutians study are made by state figures called priests within the Aleutian culture. These records are the Aleutian map of the culturally intelligible, filtered through the state to allow only what is acceptable to form the future self: 'What went on your record was your life as they saw it: your experience filtered through the state religion' (*North Wind* 33). That Jones wants us to see connections between the Aleutian practice and our own culture is suggested by the limit she provides to the material that halfcastes may use to 'find' themselves, only the records of the twentieth century, since 'further back, there weren't any moving images' and 'further forward, you ran into a modern "deadworld" tech, which the Aleutians spurned' (*North Wind* 85). This construction confronts the twentieth-century readers of Jones's fiction with our own investments in fictional characters or cultural ideals that we emulate and model ourselves upon.

Jones continually emphasizes that she wants her human readers to understand the Aleutian practice of finding oneself in cultural records to be an experience shared between humans and Aleutians. Apparent differences resolve into identity time and again. The hybrid character Catherine describes the Common Tongue as an Althusserian call:

> My mind turns it into words. I hear voices. Most Signifiers [people who speak aloud in words] do – and not only when the supposed speakers are present. But so do you, or one voice, at least. I think the interior life of an Aleutian Signifier is like the interior monologue of human consciousness, the voice that you hear in your head constantly and you can scarcely stifle if you try. With us that voice is modulated. All the possible selves of Aleutia talk to us, and we talk back to them. It's our way of experiencing social pressure, personal complexity, cultural assumptions, and so on. (*Phoenix Café* 84)

The Common Tongue is the limit culture places on the expression of the self, and the evidence that culture constitutes its subjects. Jane Ussher's article 'Framing the Sexual "Other"' suggests that humans may share this process of learning to be oneself through the study of records on a more conscious level as well. Ussher reports on the results of a series of interviews she conducted with lesbians about the process of becoming or discovering oneself to be a lesbian. The interviewees reported experiences that are analogous to the Aleutian cultural practice. Most interviewees felt that their lesbian identity was something that was always a part of them, an expression of a pre-existing interior essence rather than a conscious or unconscious sexual choice that was added to a previously non-gendered identity. However, the interviewees commonly reported that this interior essence was experienced as a vague, enigmatic desire, a sense that they were 'different' which lacked a concrete specificity. This mystery was resolved for the interviewees through their experience of consuming cultural representations of lesbianism from a variety of sources: popular productions, medical literature, and discourse circulating within their social milieu. As with the Aleutians, these individuals experienced a moment of recognition that helped them connect their interior desire with a model of how, concretely, to live that desire as a culturally intelligible identity.

Ussher's article provides a concrete example of how ideology works to shape our identities and how it works to select, in Judith Butler's terms, which bodies and selves can mat(t)er(ialize). For the interviewed subjects, 'being a lesbian' encompasses more than simply being in a female body and feeling sexual desire for other female bodies. It also means being a social subject who inhabits a recognizable space in the cultural spectrum. The 'finding' of oneself through cultural representations has both negative and positive effects. On the one hand, it provides the reassuring sense that one is not alone; there are other people who feel the same desires. On the other hand, it means coming to terms with the various assumptions and presumptions about what a lesbian identity means to this culture. In a context in which racist, sexist, and homophobic representations circulate, 'recognizing' oneself in discourse is a slippery affair. It can mean negotiating a sense of guilt that may be internalized from representations of lesbianism as 'unnatural' or 'inferior'; it can mean feeling like an inadequate subject because one does not recognize oneself in all the aspects of the lesbian identity that circulates in culture; and it can mean attempting to restructure and refigure the cultural representations of lesbianism that are available. The space for agency and resistance comes from the need to continually reconfirm one's identity through repeated performance. Interpellation is never complete or finished.

Jones's novels suggest that we also need to recognize and repeat a culturally constructed identity of what it means to be human. In each of these cases, assuming one identity means repudiating another: one becomes a man by repressing femininity; one becomes heterosexual by rejecting homosexual desire; one becomes human by refusing to acknowledge the continuity between humans and others. De Lauretis's notion of reverse discourse as resignification and rearticulation can also be extended from the subject's self-representations to its textual productions. The Aleutian trilogy may be read as a resignification of what it means to be human: the Aleutians are a representation of how humans could potentially be different.

*Phoenix Café* most clearly articulates the central theme of the trilogy that the search for a unique human identity is really obscuring an underlying unity between aliens and humans. In attempting to explain differences between human and alien culture, the Aleutian character Catherine emphasizes that they are rooted in different degrees of awareness rather than in different circumstances. She argues that societies are patterns and that 'each individual is a nexus of relationships in that pattern: a particular knot in the web that returns like a ripple in the stream, though the water is not the same water' (*Phoenix Café* 225). Aleutians are aware of this pattern and its repetitive nature, while humans are not. Thus, the Aleutian sense that they become the same person generation after generation compared to the human sense that unique individuals live only once is a matter of philosophy, not biology. She argues that humans say 'the place X is temporarily occupied by Catherine' while Aleutians say 'the place X is a person called Catherine' but that ultimately 'either explanation will do. Aleutian physiology is different from human physiology, Aleutian reproduction different from human reproduction. But our subjectivity is the same: it's the same sense of self' (*Phoenix Café* 225).

In this speech, Catherine argues for the understanding that human identity and Aleutian identity are identical: that is, they both share a common subjectivity and it is merely their social institutions and practices that give an illusory mark of difference. Catherine's analysis also points to another motif that runs throughout the trilogy: the separation of self and other. Aleutians do not recognize a separation between self and not-self (world) in the way that humans do. Jones expresses this motif of the unity of self and world in two ways: through the description of the biochemical aspect of the Common Tongue and through the search for an FTL travel device.

The biochemical communicative aspect of the Common Tongue is accomplished through ingestion of wanderers, the semi-sentient cells that are passed among Aleutians. The wanderers are not only a method of communication among the Aleutians, but are also an embodiment, literally, of the Aleutian ideas about self and world. For an Aleutian, everyone who is part of the same brood is part of the same biochemical life. All Aleutians are aspects of the same WorldSelf: all of their technology is a biological secretion of their selves, also identical. In Aleutia, 'your identity was never in doubt, it filled the air around you. ... But on earth everyone had to have a fixed title, local style' (*North Wind* 9). For Aleutians, their wanderers are both a communication of their unique identity and also an expression of identity or sameness with the rest of the world, since these wanderers can be exchanged. All Aleutian technology is also a part of this same WorldSelf, a biochemical excretion of self that becomes tool.

Organic technology is the most threatening and most alien aspect of the Aleutians from the human point of view. The Aleutian biotechnology is superior to and easily surpasses the 'dead' human technology of electrons. Humans are characterized by their need to separate themselves from the world while the Aleutians live 'in a broth of shed cells, tastes, and smells that kept them always in contact with each other' (*Phoenix Café* 18). The Aleutians, then, are the image of an ideal often represented in anti-sexist and anti-racist writing: a culture that is not rooted in hierarchies of male over female, white over black, man over nature. Instead, they are a utopian culture of harmony, offering the promise of difference without domination. Like Donna Haraway's cyborg, they eschew the boundaries of man/animal or man/machine: both their technology and their domestic animals are made from their own biochemical secretion.[9] The different sense of self that Aleutians have was designed, Jones tells us, as 'the opposite of Cartesians' ('Aliens' 114). Because human identity, rooted in Cartesian separation of mind and body, needs to distinguish self from other and to maintain boundaries between the human and not-human, the human characters in the novels are very threatened by the aliens and their technology. After the aliens arrive, humanity abandons a type of technology called blue clay, which is an organic protein technology that 'could rebuild itself in situ if strange impulses came along that needed different pathways' (*White Queen* 73), because they fear the aliens' ability to control this technology. The humans fear and loathe the breakdown of boundaries: 'Everything was alive: rock, metal, food, tools. Everything was crawling with the infection of Aleutia: a world of flesh infested with the life of its people' (*White Queen* 249).

Perhaps most threatening to the humans is the Aleutians' fluid sense of the boundary between self and other. Part of the function of the wanderers is to convey the feelings and reactions of Aleutians to one another. Aleutians do not understand privacy as a need to hide aspects of themselves from others of their brood. They continually exchange wanderers, consuming the cells of the other and thereby incorporating the other into self: 'When an Aleutian takes a wanderer from his skin, and feeds it to a friend, he's saying *this is me now, this is the state of my being*' (*Phoenix Café* 134). This exchange is increased in volume during their sexual encounters, the complete sharing of oneself with the other, and the incorporation of the other into oneself. The human response to this practice is to feel repelled and invaded. During his sexual encounter with Clavel, Johnny responds with visceral disgust: 'The truth was too vile. Things crawled, alive inside him. It was the filthiest nightmare, and it was real. He thought he would never again be free of this awareness of squirming life: on every surface, inner, outer, everything he touched' (*White Queen* 218).[10]

Johnny's rape is an extremely important event in the first novel, one whose repercussions resonate throughout the remainder of the trilogy. On the one hand, we must understand the rape as a 'real' event, given the damage that Johnny clearly suffers from this incident. From the human point of view, he has been forced to engage in intimate contact against his will, an act of violation of his self. Although part of Johnny's horror is linked to his fear of connection with all forms of life, another part has to do with the power struggle that underwrites all rapes: forcing the other to become an object. The fact that rape is an act of violence against the other's very construct of the world is made all the more apparent in the Aleutian context, given that the sharing of wanderers is literally an ingestion of the other's state of being at the moment.

From another point of view, however, we need to mitigate our sense of outrage at Johnny's suffering and see this event in the continuum of other events that mark the fundamental misunderstanding between humans and Aleutians. When the incident occurs, Clavel believes that he has found his destined love object in Johnny, his 'true parent,' which is a version of his own biochemical identity born to another generation. Aleutian romance fantasies are based on the perfect love between a true parent and true child. When Clavel approaches Johnny, they are speaking only in the Common Tongue, although Johnny is still under the impression that the communication is accomplished through telepathy and is therefore unaware of what he might be saying in Common Tongue. Johnny believes that Clavel can save him from the QV virus, an infectious virus that has been passed from technology to humans through the medium of the protein-based blue clay technology. Johnny's infection with the virus has prevented him from seeing his daughter, Bella, and therefore when Johnny expresses a desire for what Clavel is offering, believing this offer to be a cure for QV, it is not unreasonable to suppose that the desire to be reunited with his child is part of Johnny's Common Tongue expression. From Clavel's point of view, he is the true child and Johnny's receptiveness indicates desire for their sexual union.

In the conversation between Johnny and Clavel before the incident, Clavel tells Johnny, '<I'm the one who can give you what you most desire in the world>' (*White Queen* 195). Johnny believes that Clavel is referring to the cure and is 'shocked at having his need stated so bluntly' (195). Aloud, he asks, 'you can help me?' to which Clavel replies in Common Tongue '<I am as ready as I can be, without a sign from you>' (196). At this point, Johnny does say 'yes,' although he believes he is agreeing to something else. The fact that Johnny has given his consent from Clavel's point of view clearly limits our ability to read this incident straightforwardly as rape. As well, although Johnny clearly never intended to indicate that he wanted sex with Clavel, he has wanted to be able to have unprotected sex with another throughout the novel, something he is no longer able to do because he has QV. Johnny's desire to be cured is also a desire to end his sexual and social isolation.

This encounter illustrates the dangers of human interpersonal relationships and how easily we can slip between the intersubjective space of shared subjects and construct the other as an object within our own fantasy. In an earlier section of the novel, Johnny feels lust for Braemar and imagines himself touching her and removing her clothes, and then imagines that she wants him to do this. He stops himself mid-fantasy, noting 'this is how one becomes a rapist' (*White Queen* 47). Later, Johnny and Braemar do have consensual sex. After his experience with Clavel, he recalls this night and the thin line between rape and 'the best sex there could ever be in the world. When it starts off so desperate, so *unstoppable* it is indistinguishable from rape. And then you find she wanted it all the time ...' (*White Queen* 203, ellipsis in original). Realizing that he did respond physically to the sex, Johnny feels unable to charge Clavel with rape; he more easily identifies with the active participant who finds out after that the object of his desire 'wanted it all the time.'

When Braemar later talks to Clavel about the incident, Clavel has difficulty understanding Johnny's complaint that 'the worst thing was I tried to *treat him like a woman*' (*White Queen* 211). Johnny's response to the rape is to become militantly anti-Aleutian, wanting to stamp out the culture that had reduced him to feminine, passive, helpless status. He initially begins to attack Braemar, although he collapses emotionally before raping her. He recalls the rape as a 'hideous humiliation' that had contaminated him, likening the way he was constructed in the encounter to Braemar's early life as an abused wife.[11] The rape does cause legitimate injury to Johnny, but I would also argue that the human need to create a distance between self and other and a hierarchy between male and female underlies the severity of his response. Given his construction of himself as active, autonomous, impenetrable male, he is destroyed by the incident which basically destroys his self. Finally, the idea that we should understand Johnny's response as a limitation of humans and their need to construct difference and power hierarchies, especially through gender, is suggested by the fact that a later incarnation of Clavel, Catherine, lives as a human female as an appropriate form of penance for the rape.

As with the ability to use the Common Tongue, the conception of self as separate from the world or as part of the world seems to function as a marker of identity, a division between human and alien. The Aleutians are comfortable with their open borders, the interpenetration of self and other through the exchange of wanderers. They are also comfortable with and considerate of the semi-sentient 'machines' (commensals), their organic technology. They treat the commensals as fellow subjects, not as objects to be manipulated and used. The humans, in contrast, react with visceral disgust, fear, and panic at 'the reality of compatibility sans frontiers' (*North Wind* 260). However, Jones again subverts this marker and changes it into an argument for sameness, a suggestion that humans, too, are capable, as Donna Haraway argues, of seeing animals and machines as collective actors in the social construction of our reality rather than as objects to be acted upon by our subjective will.[12] Jones does this through the debate on FTL travel that extends through the trilogy.

A human, Peenemünde Buonarotti, invents a type of FTL travel in the first novel. The basis for this travel is not finding a way to move matter at a speed that is faster than light, but instead to transfer only information over the distance. Arguing that consciousness is simply an arrangement of information, she develops a technology that can transfer your consciousness to another location, providing that you can imagine the location. At the terminal point, a new body – identical to your original body – is constructed from the materials at hand. When you wish to return, an act of will can return the information – your consciousness – back to the originary point and your original body will be reassembled.

Buonarotti hides her invention in *White Queen* because she believes that humans cannot use the technology. The plot of *North Wind* is largely devoted to a search for this lost technology. The device is finally found along with a recording Buonarotti has left, part of which explains why she felt that only Aleutians are capable of exploiting the invention.

> For human beings, the experience is too much like a dream. Your mind/ brain will enact meaning on what happens, as it does on the images that pass through your consciousness in sleep. *It is impossible for a human being to take action in the visited world without falling into a psychotic episode.* The dream becomes a nightmare, in which the traveler is trapped. I have found no way out of this impasse, and because of the way we construe our consciousness – the mind in the machine – I am not hopeful that a way can be found. We humans may travel only as ghosts, shadows, spectators … If you are an Aleutian, as I believe you are, the case is different. It is the pattern of consciousness that 'travels.' For you, the pattern of consciousness is diffused through your air, your tools, your whole world. You, I believe, may find a way. (*North Wind* 275)

It is the human need to separate the self from the world that restricts humans from FTL travel. Notice, however, that Buonarotti doesn't argue that it is the *nature* of human consciousness that creates this barrier; rather, it is 'the way we construe our consciousness.' Buonarotti had earlier[13] argued that '*as far as I look into what it means to be conscious, I find an act of separation ... Consciousness is that displacement. To be unreal. To be separate from reality!*' (*Phoenix Café* 29). The heritage of Cartesian mind/ body dualism is evident in this conception of consciousness: the separation of mind from body is the act of displacement which then allows us to construe our consciousness as the distance between ourselves and the material world. Such an understanding of consciousness does not require humans to take responsibility for the worlds we make and the ways in which we treat others in these worlds.

In the final novel of the trilogy, *Phoenix Café*, the Aleutians are at work developing Buonarotti's technology so that it can be used to transport the entire Aleutian ship home. It is commonly accepted by humans and Aleutians that humans are incapable of using the technology as Aleutians do, although the humans hope to modify it for their own use after the Aleutians leave. At the end of the novel, a group of young humans who are part of the Renaissance movement – the revival of human culture and craft – reveal that they have perfected the technology for human use. The trick is adapting technology that had been used to create virtual reality games, a technology which over the centuries has produced in habitual gamers 'different neuronal mapping' (*Phoenix Café* 336) from that of 'normal' humans. Gamers are accustomed to acknowledging that the worlds they move through virtually are worlds constructed by human consciousness and perception. The technology of virtual reality combines the agency of human will (the programmers) and the constitutive aspect of human perception (the world is made through our perception of it). In the virtual reality games depicted in Jones's novels, the subject and the world shape one another.

In Jones's FTL technology, the ability to perceive the world as a construct of human consciousness is what allows humans access to other 'real' worlds. This slippage between real and virtual worlds encourages the reader to consider the ways in which the 'real' social world is a construction, a product of human consciousness and perception, and also how human consciousness and perception are products of this social world. Consciousness does not have to be an 'act of separation,' as in Cartesian theory. Even if consciousness is information, it is information that requires a material embodiment: the consciousness transported by FTL travel must materialize a body out of the world it moves into upon arrival. Outside of material embodiment, being is nothingness. Once again, the apparent distinction between humans and aliens is reduced to identity. As Catherine argued in discussing the concept of immortality, 'our subjectivity is the same: it's the same sense of self.' Although we do not have the ability to manufacture biotechnology from our bodies, Jones is suggesting that this need not be an impediment to seeing ourselves as part of the world rather than as subjects who own or control it.

In our current cultural context, our fear of fluid boundaries between self and other is most strongly expressed in what Corrine Squire has called 'AIDS Panic.' Because AIDS is transferred through the exchange of bodily fluids, and because AIDS is fatal, it is perhaps not surprising that the image of the body as a closed system is appealing. However, as Squire has pointed out, our fears about AIDS are not simply a rational response to a communicable disease. Instead, they tie into pre-existing fears and hatreds of other-sexed or other-raced bodies.[14] This notion that humanity is naturally divided up into pure bodies and polluted bodies distorts discursive representations of AIDS. Squire sees this intersection at the root of social responses to AIDS which focus on identifying and monitoring at-risk activities and at-risk groups rather than representing AIDS as a general public health crisis. She argues, 'AIDS science is itself heavily infected by the patriarchal, heterosexist, racist assumptions of the language that writes it, assumptions that lead to partialities or omissions, apocalyptic warnings, melodramatic over- or under-statements' (52). Such distortions, she warns, impede accurate public education and direct research resources to ideologically sanctioned investigation only. Squire's analysis suggests one of the consequences of the human (from Jones's point of view) conception of life as a series of separate and discrete organisms rather than seeing all things as continuous with the self. The human perception stigmatizes and isolates individuals and refuses the perspective that there is a connection among all humans, as well as between humans and the rest of the world. Resources committed to curing AIDS are not resources wasted on an infected 'them' at the expense of an innocent 'us.'

One of the ways in which this discourse circulates is in representations of AIDS as punishment for 'sinful' behaviour such as homosexuality or drug use. The Aleutian trilogy critiques this notion of viruses as ideologically discriminatory through its representation of the Aleutian technology of weapons. All Aleutian technology, including weapons, is biological. The distinction between weapons and other products is that weapons are created from inert, non-sentient material while other products, the commensals, are created from living tissue. The destructive capacity of weapons is integrally linked to the Aleutian sense that all members of a brood are composed of the same flesh, the same self. Weapons are organic, constructed from enemies' flesh and dedicated to destroying everything they encounter that is of that flesh: Aleutian weapons refuse the possibility of sorting ally from enemy, self from other. They 'attack and consume anything that shares biochemical self with the enemy. People, living machines, buildings, food plants, the microscopic traffic in the air,' and they cannot be destroyed by any known defence. Weapons 'go on until there is no food for them left. No people, no commensals, no plants, no tools, nothing. It is genocide. There is no other outcome' (*Phoenix Café* 269).

In the final novel of the trilogy, humans attempt to make their own version of Aleutian weapons, for use in the Gender War. This is a war between two ideological positions – Traditionalist and Reformer – that is popularly understood as the war between the Men and the Women.[15] Catherine tries to point out that the notion of distinguishing self from other in an Aleutian weapon is 'insane! Weapons attack biochemical identity. They can't distinguish between *political parties!*' (*Phoenix Café* 298). Clearly, the notion that weapons designed to attack based on biology would be able to distinguish between political parties is illogical. This incident reveals the absurdity of our cultural constructions of identity based on the body and our social discriminations between different bodies. Is it any less strange to think that we can assess someone's worth, their abilities and potential, based on their gender or their race? Is there something fundamentally different about the races and the genders that biochemical weapons could discern?

The appeal of such ideas to certain humans is apparent in continued attempts to locate and define differences between sexes, races, and sexual preferences through a reading of the body's chemical or genetic information. In Catherine's speech, we see again the message, often repeated through this trilogy, that humans and the rest of life on earth are one WorldSelf. The risk of deploying Aleutian-style weapons is the risk of genocide, not just for all humans on earth but for all life on earth. These associations link Jones's description of how humans could be different – how we can understand her fictional Aleutians as a model for how we could construct our understanding of humanity – to environmental and anti-nuclear discourses. The models we construct to understand humanity and its relation to the rest of the world construct meaning or a version of the 'facts'; they do not simply describe the 'reality' we find.

I now want to turn to considering how the definition of identity as the individual characteristics by which a person or thing is recognized circulates through these novels. One of the main questions raised by the novels is: what is cultural identity? As I have already argued, the novels portray the attempts of human characters to establish the boundaries of human cultural identity and the deconstruction of these attempts to root individual identity in difference from the other. As well, the representations of cultural identity in the novels reject mind/body dualism. I now want to turn to the novels' interrogation of the culture/nature binary in assessing the relative impact of biology and ideology on the construction of identity. This interrogation is accomplished through the characters of Bella in *North Wind*, and Catherine in *Phoenix Café*. While the halfcastes are figures of cultural miscegenation, Bella and Catherine represent a type of biological miscegenation, although they are actually more akin to the transgenic products of modern genetics than they are to the offspring of different races.

Bella is the product of a tissue sample taken from Johnny Guglioni, which somehow has been modified to produce an Aleutian fetus.[16] The Aleutians have two motives in producing this hybrid: to heal an interpersonal conflict and to reveal the location of the FTL travel device. Both goals would require that Bella access Johnny's memories.

Although Bella's body is Aleutian in its anatomy – no nose, Aleutian hermaphroditic genitals, limbs which can reverse in a way that human limbs cannot – her[17] biochemistry seems to be human. Bella's true identity is not revealed until near the end of the novel. For most of the novel, she is presumed to be a person from the Aleutian ensemble who has a disability, no wanderers, and no ability to perceive and exude bio-chemical signals of self. This disability puts Bella outside of Aleutian culture in a number of ways: she is impeded in her ability to read the Common Tongue; her sexual encounters cannot include the sharing of wanderers; and she can recover only a part of the lesson provided by studying records of her previous lives, as the records exude biochemical information that is inaccessible to Bella. Bella's lack of biochemical dif-fusions means that, unlike most Aleutians, she can be confused about her identity, which is the central mystery of the novel. At various times in the novel she believes that she is Maitri's librarian, the identity she has studied in the records; Johnny's human daughter, also named Bella, who she believes was kidnapped from earth by Aleutians; and finally – when her origins are revealed – the reincarnation of Johnny himself. When Bella discovers the truth of her creation, she complains, 'What had been done to her was cruel. If you don't know who you are, you are cut off from the WorldSelf. You can't know God if you don't know what aspect of God, the WorldSelf, is you' (*North Wind* 204). She withdraws from Aleutian culture for a time and lives on earth among the halfcastes. Ironically, she is the only biological hybrid among them.

Biological manipulation is not able to bridge the two cultures; Bella does not have Johnny's memories and cannot heal the past between Johnny and Clavel or lead the Aleutians to the FTL device. Bella ulti-mately comes to the conclusion that 'Race is bullshit, culture is every-thing. No matter how I was built, I'm an Aleutian' (*North Wind* 250). The representation of Bella thus supports the nurture side of the nature/nurture divide. Of the two components that work to form a cul-tural identity – biological material or DNA and studying the records or ideological interpellation – it is the cultural interpellation that forms Bella's self. Such a representation cautions against ideologies that sug-gest that the body is a reliable informant that can be used to discover the 'truth' of the subject; the nature of the body does not exist outside of the nurture that structures the meaning of its various activities and expressions. Bella's body is Aleutian because her relationship to it has been formed within Aleutian ideology, as is evidenced by her feeling that she has been cut off from the WorldSelf by having her Aleutian identity hidden from her.

In contrast to Bella, human DNA in Aleutian body, Catherine is the product of Aleutian DNA formed into a human body. The current rein-carnation of Clavel, Catherine has chosen to spend his final days on earth as a human female. Like Bella, Catherine is the product of genetic engineering: an Aleutian fetus which is transferred into a human womb and modified so that its morphology will be human. Catherine's anat-omy is human but her memories of her previous lives, as Clavel, cause her to sometimes feel uncomfortable in her body. Its actions and limits are strange and unknown to her: 'The body was human, the spirit knew a different set of rules' (*Phoenix Café* 18). Although she is limited by her human body, Catherine has no doubt about her identity: 'The records had a biochemical content that her human body could not process: a haze of living inscription that left the screen but could not penetrate her human skin. But one day it had come to her, exactly as if she were a normal Aleutian, without a shadow of doubt, that this was herself. *I am me*' (*Phoenix Café* 24).

The combination of Bella and Catherine reinforces the idea that 'race is bullshit; culture is everything.' Although Catherine is born from a human mother, Maitri raises her as his ward, entirely within Aleutian cultural norms. Her birth mother will not acknowledge that Catherine is her daughter in any way, and always calls her 'Miss.' Although Catherine has human DNA, she is no more capable of understanding the human culture than any other Aleutian. Having a human body does not give her a human personality, and having a female body does not make her a 'woman.'

The representations of Bella and Catherine suggest that different bodies do not guarantee different identities. The relationship of iden-tity to the body is important to understanding the trilogy because of the social differences that humans construct about morphological dif-ferences. This idea is expressed in the novels through the halfcaste characters, and through the Gender Wars which form the background to the alien encounter. If the sense of self diffused through the entire world is the most difficult aspect of the Aleutians for the humans to comprehend, gender division is certainly the most difficult aspect of human culture for aliens to grasp. The aliens insist upon seeing the humans as a single brood – all made of the same Self from the Aleutian point of view – and they cannot understand the human division of the species into two groups based on morphology. They grasp that the humans are divided into political factions which seem to be rooted in gender distinctions, a concept which is partially fuelled by the fact that they mistake the World Conference on Women which is in session when they arrive for the world government. The Aleutians feel that their inability to differentiate the two broods causes them to make mistakes in local politics. By the conclusion of *White Queen*, the aliens have requested that humans wear uniforms in order to specify gender when interacting with Aleutians: the Woman uniform has padding to emphasize the breasts and buttocks and the Man uniform has a well-padded codpiece.

According to a human character in the novel, the gender conflicts being discussed at the World Conference on Women are really issues of labour. The combination of patriarchy and capitalism has produced a situation where the most disadvantaged economic class and those who suffer the poorest working conditions also happen to be females. The conflict escalates into a series of worldwide riots, called the Eve Riots by the media. From that point forward, the distinction between gender conflict and political conflict becomes blurred for both humans and Aleutians:

> Thereupon a whole package of worthy, virtuous reforms of human behaviour, including a better deal for 'biological females,' became known as the Women's Agenda. After the sabotage crisis there was a backlash, and the Men's Agenda emerged. The Men were the traditionalists, and they included plenty of women who agreed that a traditional division of labor, responsibility and material wealth between the genders was natural and right. (*North Wind* 23–4)

The human notion of gender has been confusing to the Aleutians from the start. The explanation for the division that circulates among the Aleutians is that the human race is divided into two broods, which they refer to as the 'childbearers' and the 'parasites.' As the conflict shifts from the gender-based discussion started at the World Conference on Women to the more generalized political conflict between the Traditionalists and the Reformers, the Aleutians become increasingly confused about human gender. The problem is that 'there are biological males on the Women's side, and biological females on the Men's side' (*North Wind* 17).

The Aleutians are unable to comprehend the connection some humans draw between biology and gender because while the Aleutians do have a concept of gender, their divisions are rooted in personality differences, not bodily morphology. Further, they struggle to comprehend the centrality of gender difference in human life, given that 'in Aleutia worrying if you were "masculine" or "feminine" was the sign of a trivial mind' (*North Wind* 28). From this Aleutian point of view, the 'natural' connections that are often assumed between biological sex and social gender appear as arbitrary constructions. As Judith Butler has argued, gender is a performance that demonstrates an allegiance to a particular ideological picture of reality. In the world of these novels, this relationship is made very clear: adopting gender norms indicates support of the traditionalist's Men's agenda, while resisting gender norms indicates support of the reformer's Women's agenda. Whether one attempts to essentialize gender in terms of morphology, as the humans do, or personality, as the Aleutians do, both systems are explanations that reveal the 'natural' to be ideological, only ever a partial explanation.

Judith Butler analyses a heterosexual norm and the ways in which both desire and personality traits are expected to follow 'naturally' from biological sex. The Aleutian trilogy does not explicitly take up this issue. Aleutian cultural practice separates sexual activity from reproductive activity. This suggests that, from their point of view, the notion that desire is a natural expression of biological morphology would also be incomprehensible. Jones is able to critique our sex/gender system by showing the reader how ridiculous and artificial it is.[18] She also makes many ironic comments about the destruction caused by this system. During the three-hundred-year period that comprises the novels' setting, human social conditions deteriorate, largely due to the contamination of land so that it is no longer suitable to produce crops. Many humans blame the Aleutians for this problem, since the Aleutians have established some terraforming projects that have had negative side effects. The novels clearly state, however, that the destruction is mainly the result of the continuing gender wars. Despite the arrival of a species from another planet – and, as the humans gradually realize, a hermaphroditic species at that – humans continue to tear apart their world over conflicts about gender roles. Jones suggests that some of the blame may be laid at the aliens' door, simply because they are hermaphrodites: 'The superbeings made it valid for everybody to be a person. But – cut it any way you like – that means there's twice as many fullsized humans in any given area than there used to be, and still only one planet. Naturally, there's a war' (*North Wind* 95). The bottom line is that argument about gender, or race or any other feature of the body as a marker of social worth, is ultimately argument about the appropriate distribution of material resources.

Again and again in these novels, Jones shows humans clinging to the notion of a natural body and then deconstructs these representations to reveal the body as a social product. Halfcastes mutilate or, less pejoratively, modify their bodies so that they resemble the gender-neutral and noseless Aleutians. The halfcaste culture, on the surface, supports a reading of the body as necessary marker of group; in order to be like the Aleutians, the halfcastes must modify their bodies. What the representation of the halfcastes ultimately suggests, however, is that the shape of the body is irrelevant. The body modification does not bring the halfcastes any closer to an Aleutian identity – in fact, the Aleutians are rather suspicious of halfcastes. It is only the humans who take this body modification seriously, subjecting the halfcastes to various discriminations ranging from social slights to pogroms. The halfcaste culture shows that gender is a performance that does not have to be enacted, regardless of physical anatomy. In fact, halfcaste culture demonstrates the degree to which all identity is an imitative performance of ideological norms. A key halfcaste character in *North Wind* explains, 'I'm not a Man, not a Woman, not an Aleutian. Whichever role I take, I feel like one of the others playing a part. That's why I call myself Sidney Carton. He's a character in a drama-movie, chiefly famous for pretending to be someone else' (*North Wind* 19). The halfcaste culture expresses the link that may be drawn between Aleutian cultural practices and the ideological world outside the novel. The cultural representations we consume are our imagined selves, even if we do not institutionalize this assumption of subjective images as a religion.

In a discussion of representations of the body as marker of social meaning, it becomes apparent that one of the things that is absent from these novels is an explicit representation of the discourse of racism. The novels do present human characters of many different races, and the action is set in many locations including Africa and Thailand as well as Western locations. In 'Yellow, Black, Metal and Tentacled' Edward James has argued that the question of race is often displaced onto the alien in science fiction. What this often means is that the world of the science fictional text is peopled entirely by white people, and the text's anxiety about other races is transferred to the aliens in the text. While there are non-white characters in this trilogy, there are no representations of racial tension between them. However, the anxiety and hatred often expressed by racist discourse is found in human responses to the aliens. In *White Queen*, the eponymous anti-Aleutian group attempts to sabotage the alien ship; in *North Wind*, an anti-Aleutian uprising forces the aliens to leave earth for a time; and in *Phoenix Café*, the Renaissance movement is committed to recovering a lost human technology and culture, seeing the human way as essentially better than Aleutian or hybrid practices.

Although there are no representations of racial conflicts between humans, race, then, is not an entirely absent discourse. The representations of resistance to Aleutian presence suggest the alien encounter as a metaphor for colonial encounters. The language of Indian colonialism is used in exchanges between Aleutians and humans in *North Wind*: Sahib and 'tame' locals. As with many historical colonial encounters, it is technology that most distinguishes the Aleutians from the humans. Braemar Wilson, the leader of the White Queen, an underground organization opposing the aliens, explains her antipathy toward the Aleutians in terms of fear of human obsolescence: 'I know that they don't mean us any harm. They don't have to. It's historically inevitable. If they're superior that means we're inferior. Right? If we're inferior, then give it a generation or so and there'll be Aleutians in our jobs, Aleutians in the White House; and two doors to every leisure center; two kinds of life' (*White Queen* 178). Braemar's assessment, which seems excessively paranoid in the first novel, has proven to have some basis by the third. The Aleutians have not taken over human political structures; they have simply replaced them, dividing the world up into protectorates in a style reminiscent of the British Empire. The Aleutians' purpose has always been economic gain, a final and extremely telling colonial link.

Johnny's rape becomes a metaphor for understanding the relationship between the colonizer and colonized, the exploitation of one party to satisfy the other party's desire. For Johnny, one of the worst aspects of the experience was his own seeming complicity. Although he did not understand himself to be asking for sexual contact, he found that he was aroused by the experience nonetheless. Johnny feels most violated by the sense of having his own body turned against his conscious desires, feeling torn between his sense of having been raped and his sense of having been sexually gratified. This sense of doubled consciousness is the aspect of rape most useful for understanding it as a metaphor for the colonial experience. The struggle between the colonizer and the colonized is ultimately a struggle about the meaning of events. Their 'truth' is produced through how they are represented, as riot or as protest, as war of liberation or as rebel uprising. The metaphor of rape explains this struggle over imposing one's ideology, one's truth, on the other:

To an Aleutian rape has the same meaning – if you can call it that – as the human act. When they *lie down* together, as they say, the exchange of wanderers, a polite constant in social interaction, becomes a flood. Wanderers are directly absorbed through mucous membranes, sometimes in enormous numbers. Thus, their lovemaking, as the human gestalt, is essentially an act of chemical communication. Rape, as among humans, is the means of imposing a stronger party's version of events on futurity. (*North Wind* 94)

This particular rape further resonates with colonial discourse in the sense that Clavel thought that he was responding to Johnny's desires, helping Johnny as it were. Versions of imperialism have been justified by this discourse about 'helping' the other.

By the third novel, the natural resources of earth have been exploited or destroyed – granted, this is not the aliens' fault – and the Aleutians are preparing to depart the depleted planet and leave the humans to struggle through rebuilding. The Aleutians have radically altered the earth in ways both damaging and useful. Their project to flatten the Himalayas, a long-term climate improvement project, reduced the amount of arable land and caused many (temporary from the Aleutian point of view) deaths. On the other hand, their biological technology eliminates many pollutants from the water supply and provides a method for preparing uncontaminated food without requiring clean surfaces. Jones presents a very ambivalent attitude regarding the consequences of the encounter with the Aleutians, 'ignorant, well-meaning foreigners' whose technology has 'made an immense difference to the poor; far more difference, more quickly than we could have made with political solutions' (*North Wind* 107). It is possible to read this as a defence of colonialism: yes, mistakes were made but superior technology ultimately improved the lives of the natives. However, I believe that the novels suggest something else, an idea related to how the alien encounter is typically presented in science fiction.

A familiar trope in science fiction is that of the alien as superbeing whose superior knowledge and/or superior social organization will save human beings from our own destructive tendencies.[19] Jones makes it quite clear that the humans in the novel are responding to the alien presence through the expectations that have been generated by the tropes of the genre. Describing the education in xeno-anthropology provided to those humans chosen to interact first with the aliens, a representative reports, 'We were shut up for a weekend ... And forced to peruse an inordinate quantity of science fiction' (*White Queen* 97). This

comment, humour notwithstanding, points to the most important argument that Jones is making through this set of texts. The aliens will not
save us. We must fix the damage we do to ourselves, our constructed
others, and our world. We must discover ways to work through conflicts
created by discriminatory constructions of race, gender or other difference. We must fix the damage we do to our environment that inevitably
threatens our own survival. We must fix problems of overpopulation,
starvation, lack of potable water, poverty. In short, the trilogy is about
how we must find our own solutions to our own problems. It explicitly
refuses the rapture-like escapist ideology found in many other science
fiction texts, such as Clarke's *Childhood's End*, Bear's *Blood Music*, or the
recent *Left Behind* series, by Tim LaHaye and Jerry Jenkins, which tells
the story of the biblical rapture as science fiction.

The emphasis throughout the trilogy on the connections between self
and world suggests that any solutions we do find to our problems must
be collective solutions, not the individual heroic solutions also often
favoured by science fiction. My reading of this trilogy has also pointed to
the ways in which it demonstrates that concepts of the body inform our
political and social choices. The notion that mind and self are separate
from the material world of things encourages us to see ourselves as separate from nature and entitled to exploit it as an object. Mind/body dualism is a way of constructing a sense of self out of repudiation, a process
that projects all the disavowed qualities on to the body and our constructed others. The body is then associated with these others (traditionally female and non-white), while all positive qualities are associated
exclusively with the self/mind (traditionally male and white). These others are then despised and diminished as the bearers of all that is negative, fleshly, and basely material. Clearly, a new concept of the body and
of identity is required if we are to change these social structures.

While refusing this rejection of the body, Jones does not reify the
body as that which determines character or identity. In her presentation
of Aleutian culture she portrays a relationship between body and self
that could be described by Grosz's metaphor of the Möbius strip: internal and external integrally bound together. Biochemical signals and
internalization of cultural representations both form identity. Neither
the body nor culture can be ignored in our attempts to reshape ourselves, and the trilogy provides a possible model for a way to begin to
rethink human identity as part of the world, connected to the material
rather than transcending it. Jones's *Aleutian* trilogy is a reverse discourse
that rejects the ideological call of a subject formation founded on repu-

diation and the Cartesian mind/body split. Instead, she argues for a vision of ourselves as more like than unlike the aliens. At the same time, her representation of the process of interpellation through the Aleutian practices of 'becoming your self' works to make visible the operation of ideology, to undermine its ability to appear obvious.

One of the critical principles articulated by Jones in her writing about science fiction is her belief that the alien in science fiction is always a representation of 'other people.' The figure of the alien is one of the ways available for the science fiction writer to explore those we deem different from ourselves, with all the risk of repudiated identifications that this construction of difference entails. On a more hopeful note, Jones also suggests that 'Sometimes science fiction aliens represent not merely other people, but some future other people: some unexplored possibility for the human race. Maybe my Aleutians fit that description' ('Aliens' 119).

As they prepare to leave earth at the end of the final book, one of the Aleutians observes that humanity has changed through its interactions with the aliens: 'I think you no longer quite see yourselves the way you did when we arrived, as separate objects in empty space. You feel your-selves to be, like us, part of a continuum. Part of the heterarchy of life, where it's natural for all boundaries to be in continuous negotiation' (*Phoenix Café* 227). This self-in-connection is what Jones's work ulti-mately calls for us to be, a new possibility for human identity beyond repudiations and hierarchies, an inclusive rather than exclusive post-human identity.

# 2 Octavia Butler: Be(com)ing Human

The atomic age began with Hiroshima. After that no one needed to be convinced we had a problem. We are now entering the genetic age; I hope we do not need a similar demonstration.

Robert Sinsheimer, 'An Evolutionary Perspective for Genetic Engineering'

Both science fiction and discourses on genetics are concerned with marking the boundaries of humanity; Octavia Butler's *Xenogenesis* trilogy demonstrates these parallels.[1] The trilogy tells the story of a group of humans who have survived a nuclear holocaust and of the alien species who have enabled that survival. The aliens, Oankali, are a trading species and the medium they trade is genes: 'We trade the essence of ourselves. Our genetic material for yours ... We do what you would call genetic engineering. We know you had begun to do it yourselves a little, but it's foreign to you. We do it naturally. We must do it. It renews us, enables us to survive as an evolving species instead of specializing ourselves into extinction or stagnation' (*Dawn* 43).[2] The Oankali continually change themselves, evolving into new and different forms through the incorporation of DNA from the other species they encounter as they travel through space.[3] As the surviving humans discover, the Oankali do not plan to simply exchange data and part ways. They envision a partnership between humans and Oankali in which interbreeding will eventually produce a third species, neither human nor Oankali.[4]

The Oankali are a three-gendered species with 'normally' gendered males and females and a neuter-gendered third partner called an ooloi. Ooloi have a special organ – an organelle – that they use to 'perceive DNA and manipulate it precisely' (*Dawn* 43). In Oankali reproduction, there are five parents: Oankali male and female, partner-species male and female, and ooloi. The ooloi performs a role that is somewhere between that of genetic engineer and IVF technician, taking DNA from all five parents, editing and mixing it to produce the desired result, and then implanting the embryo in one of the female parents. One of the central motifs organizing the trilogy is the Oankali's contention that humans are genetically flawed. Oankali see humans as 'fatally flawed' because they have 'a mismatched pair of genetic characteristics' which 'together are lethal' (*Dawn* 40). The problem is that humans are both intelligent and hierarchical; even worse, hierarchy is 'the older and more entrenched characteristic' and 'human intelligence serve[s] it instead of guiding it ... [does] not even acknowledge it as a problem, but [takes] pride in it' which the Oankali claim is 'like ignoring cancer' (*Dawn* 41). Owing to this flaw, the Oankali deny the ability to reproduce to those humans who choose not to breed with the Oankali. Although they will allow such humans – called Resisters – to live autonomously and separately, they feel that 'it *is* a cruelty ... [to] give them the tools to create a civilization that will destroy itself as certainly as the pull of gravity will keep their new world in orbit around its sun' (*Adulthood Rites* 463). The Oankali are genetic essentialists, and they believe that the human genetic predisposition toward hierarchy inevitably dooms human civilization.

The first novel in the trilogy, *Dawn*, describes the experiences of Lilith Iyapo, a black woman whom the aliens wish to use as their primary liaison. The Oankali have taken all the humans who survived the nuclear war, put them into stasis on the Oankali ship for approximately 250 years, and are now beginning to awaken them to explain the situation. During this time, the Oankali have restored the earth so that it is again suitable for habitation. As Ruth Salvaggio has observed, Butler's heroines are characterized by their ability to compromise and survive, and Lilith is no exception. As the human who demonstrates the best ability to accept this situation, she is chosen by the Oankali to be the person who awakens the first group of human colonists and teaches them the skills they need to return to a renewed earth, one which lacks the technology that had characterized their lives before the war.[5] The second novel, *Adulthood Rites*, concerns the experiences of Lilith's Oankali-human son, Akin, and is largely devoted to his life project of defending the human Resisters, arguing that they deserve a chance to continue in a genetically unadulterated form. This novel ends with the Oankali acceptance of Akin's proposal that Mars be terraformed to support life and that the human Resisters have their fertility restored and be allowed to settle on Mars. The final novel, *Imago*, primarily concerns Lilith's Oankali-human ooloi child, Jodahs. Jodahs is the first ooloi born of the human-Oankali mix. Ooloi are the controlling participants in reproduction, since they are the ones who choose the genetic configuration of the offspring. Jodahs's birth, therefore, marks the maturity of the new genetic mixture of human and Oankali, finally able to reproduce themselves without mating with older, pure Oankali ooloi.

Scholars have typically interpreted the trilogy in terms of its comments on race relations, seeing the genetic science in the trilogy as a metaphor that allows Butler to explore human notions of hierarchy and human tendencies to evaluate others through body-based categories. Walter Benn Michaels, Amanda Boulter, Stacy Alaimo, and Donna Haraway (in *Primate Visions*) have all offered readings of the novel that emphasize its engagement with racial politics and, in the case of Alaimo and Haraway, the intersection of racial and gendered readings of the body in the presentation of Lilith. Cathy Peppers and Michelle Osherow both see the trilogy rewriting origin stories about being human and refiguring them otherwise. Rebecca Holden also reads Butler's work as a new ontology in the mode of Haraway's cyborg, refusing the category of the natural; she argues that 'Survival for the *Xenogenesis* human Cyborgs is not simply a matter of exploiting their cyborg positions but necessitates profound self-betrayal ... Making meaningful and potent connections with the truly different, Butler seems to say, will not be easy but may be necessary' (49).

I agree with Holden's emphasis on the provisional, ambivalent, and painful qualities of Butler's fiction. In my reading of the novel I want to emphasize the place of the body in Butler's cyborg politics, the visceral quality of the connections that have been undertheorized in earlier criticism. What has been overlooked in these mythologized readings of the *Xenogenesis* trilogy is the way these novels also engage with contemporary discourses about the appropriate uses of various biological technologies of body manipulation, specifically genetic engineering and in-vitro fertilization (IVF). I concede that Butler's focus in the novels is on the more abstract questions of how humans construct and value bodies, rather than more concretely on the details of how genetic manipulation might be accomplished. However, I believe that these more abstract debates about the relationships among identity, genetics, free choice, and destiny are precisely the terrain being fought over in discussions of genetic futures. Thus, Butler's work is extremely valuable for its ability to let us see the assumptions about the body that are informing contests over genetic technology. Before turning to a more detailed reading of the novel, I will review some non-fictional accounts of genetic discourse in order to make these parallels evident.

Recent advances in genetic engineering suggest that the gap between science fiction imaginings and contemporary possibilities might not be that great. Newspaper accounts promise that 'Personality can be transplanted'[6] and that smarter mice can be produced by inserting an extra gene.[7] Genetic testing and biological manipulation are increasingly becoming a part of everyday life and consumer culture. Patented genetic interventions include Microsort, a process available at some IVF clinics that allows one to choose the sex of one's fetus.[8] The cloning of Dolly and her plethora of medical problems regularly made the news. We live in a world where we can eat tomatoes that have a longer growing season because of the introduction of fish genes, and where we might soon be able to dine on the flesh of geep – a chimera or blended species produced by genetically engineering the offspring of goats and sheep. Terms such as biopiracy, bioprospecting, pharming, and humanimals have all recently been coined to describe a world in which genetic material has become just one more natural resource exploited for capitalist accumulation and in which the distinction between animals and tools is eroding. Monsanto Corporation is trying to control agriculture through the engineering of crops which are resistant to Monsanto herbicides and whose seeds are sterile, requiring farmers to return to Monsanto each planting season. We now do shape life to serve 'our' needs, but in a manner far different from the connection across species imagined by Gwyneth Jones or Octavia Butler.

Boundaries are extremely fluid in this new genetic age. Once the notion that animals are machines could be firmly understood as merely the philosophical speculation of Descartes, based more on his own projections than on animals' qualities. Now, we have produced animals as machines in a very literal sense. Certain genetically engineered animals can secrete various biochemical substances in their milk that are needed for human medical treatments; others may be genetically modified to make tissues necessary for human surgeries (Gray, *Cyborg Citizen* 122). We also create organic research machines as we patent modified animals whose germ lines have been modified to produce disease-carrying versions for research (Gray, *Cyborg Citizen* 116). The boundary between humans and animals is also eroding as we create animals to serve as organ donors in xenotransplantation work (Gray, *Cyborg Citizen* 78).

Humans, too, are or can be subject to the various protocols developed for use on other life forms. Our attitude toward the body is of paramount importance when considering how such technologies might be deployed. Our ethical engagement with these technologies lags far behind our scientific capacities. The entire human genome is now mapped, but the results of this research are owned by Celera Genomics, who profit from licensing fees and patents whenever any of this data leads to medical applications (Best and Kellner 115). The entire Human Genome Project spent only 3 to 5 per cent of its $3 billion budget on legal, ethical, and social issues, and Celera spent even less (Best and Kellner 143).

Genetic information and social conclusions based on it can have pernicious as well as beneficial consequences for social subjects. In their analysis of some of these consequences, Nelkin and Lindee argue that this particular historical moment is characterized by what they refer to as 'genetic essentialism,' an ideology which 'reduces the self to a molecular entity, equating human beings, in all their social, historical and moral complexity with their genes' (2). Whether or not one is troubled by this propensity for genetic determinism is related to how one constructs genetics in discourse. Those who favour the advances made possible by genetic engineering tend to construct a continuum between such practices and more accepted manipulations of nature such as selective crop-breeding, animal husbandry, and clinical medicine. The improvements they extol include more robust and nutritious food sources that will solve the world's food distribution problems, and human freedom from disease.[9] When considering human genetic engineering, such enthusiasts tend to construct the practice in relation to other advantages that people 'naturally' try to give their children, such as a better education. Improved health is most commonly presented as the key benefit that genetic research and engineering offer to humans. With the aid of genetic screening, fetuses with single-gene defects such as Down's Syndrome, Tay-Sachs disease, and cystic fibrosis can now be identified and aborted. Additionally, scientists believe they will soon be able to identify genetic predispositions to other diseases such as cancer or heart disease and thereby counsel the patient toward a lifestyle that will minimize his or her risk.

Those who are more pessimistic tend to construct their representations in terms of the continuum between genetics and eugenics as social practices and warn of the dangers of social discrimination that are inherent.[10] Critiques of genetic engineering point to problems in its application for health improvement. Nancy Wexler questions the ethics of providing genetic screening in cases where there is an ability to identify the disease but no ability to treat its effects (231).[11] Nelkin and Tancredi warn of the power of 'scientific information' to be interpreted as hard fact by the general public, erasing the evaluative role of the clinician. As well, they argue that our cultural mood of genetic essentialism means that whenever a disease is determined to have a genetic proponent, there is a tendency to read that proponent as the determining factor for the disease (41). Russo and Cove also identify this problem in their discussion of the search for a genetic cure for cancer even though testing suggests that 70 to 90 per cent of cancers are caused by environmental rather than genetic factors (123). Finally, as Miringoff points out, while the goal of predictive diagnosis is to counsel people to alter their lifestyles to escape the consequences of their genetic predispositions, such an option is often only available to the wealthy (68). In the absence of changes to the social structure, the benefits of genetic engineering will be restricted to the economic elite.

This economic slant is most apparent in the application of genetic screening during pregnancy and in the genetic manipulations associated with assistive reproductive technologies (ARTs). Farquhar demonstrates that the discursive image of the infertile woman – as white, middle-class, heterosexual, and married – is at odds with the real experience of infertility, which affects more coloured and poor women than white women and which is not limited to heterosexual couples.[12] To a large extent, these discursive constructions do reflect the reality of those who have access to ARTs and genetic screening during pregnancy. Such technologies are deployed along political lines and, in general, reflect a desire to solve social problems with individual, medical solutions rather than through systemic change. Arguing that more social benefits could be produced through programs such as subsidized prenatal nutrition and chemical-abuse treatment for pregnant women, Miringoff writes, 'while we utilize expensive genetic procedures for middle- and upperclass women, to prevent diseases such as Down's syndrome, we permit lower-class women to give birth to underweight, undertreated, and chemically exposed babies, producing similarly injurious forms of disease and retardation' (150).[13] Although the lure of profit has increasingly meant that those who provide ART services will offer them to anyone who can afford the price, in the past there have been attempts to restrict access to heterosexual couples. As recently as 1994, the American Fertility Association produced an ethics report arguing that 'a married heterosexual couple in a stable relationship provides the most appropriate environment for the rearing of a child' (Farquhar 181).[14]

The most prevalent and realistic fear associated with the possibilities presented by genetic engineering and genetic testing is that this data will be used to create discriminatory social categories. As discussed above, the current 'treatment' available for those fetuses identified as having a genetic disorder is abortion. In this social context, it becomes overwhelmingly important to understand what is being labelled as a disease and in whose interests such labelling is being done. Often this debate can become one about which lives are worth living and which are not. While most people would support aborting a fetus who has Tay-Sachs disease – inevitably fatal during the first few years of life – the issue becomes more complicated when considering a disease such as Huntington's, in which the afflicted individual will show symptoms only after two to four decades of 'productive' life. As many writers point out,[15] while the rhetoric is that abortion is the answer that is best for the child – to avoid its suffering – it can all too easily slip into a discussion of the relative costs to society and contributions to society that variously 'disabled' people make. The consequence is that a '"bottom line" on human value emerges' (Miringoff 18). Further, Miringoff indicates that activists for disabled rights argue against selective abortion of disabled fetuses, a perspective which tends to undermine arguments that it is for the sake of the child that the abortion is done.[16]

In addition to control over which bodies will be allowed to be born – a genetically controlled selection of which bodies matter – there are fears that genetic information will be used to discriminate against those bodies already here. Disabled people fear that social attitudes toward them will change based on the perception that they are people who should not exist, an extension of the attitude toward disabled fetuses as babies that should not be born.[17] In the United States, the overwhelming fear is that information about genetic predispositions will be used to deny medical insurance to those who most need it.[18] Both Miringoff (13) and Nelkin and Tancredi (83) cite evidence that testing for genetic predispositions is being used to restrict employment opportunities for jobs in which exposure to chemicals is part of the work environment and the person has tested positive for a susceptibility to such chemicals. As with other examples, the problem is reduced to the individual – who has a problem of susceptibility – and removed from the social – a workplace environment which should be made safer for all workers.

In each of these examples, the common theme that emerges is that social problems can be solved through biology, an attitude Miringoff calls genetic welfare and defines as 'the desire to improve the human condition through genetic and reproductive intervention. Its proponents are those who seek to modify the genetic foundation of human life, to bring about a healthier society, with fewer human problems' (6). The perspective of genetic welfare is that it is the body, alone, that determines our social being, and that the body can unproblematically be understood as something purely 'natural,' an immutable, biological given of all one's abilities and shortcomings. Changing the body can take the place of changing social structures and can eliminate everything from disease to poverty to anti-social behaviour.[19] As Nelkin and Lindee succinctly put it, DNA has become 'the secular equivalent of the Christian soul' (2). DNA is seen as holding the key to what makes us human and to what makes us the particular humans that we each individually are.

The most problematic social consequence associated with this understanding of genetics is the idea that genetics can be used to distinguish what constitutes a normal from an abnormal example of a human. The Human Genome Project is popularly represented as a search for the 'code of codes,' the 'Holy Grail,' the 'bible,' or the 'key' that will allow us to decipher humanity.[20] One of the problems with this perspective, as Evelyn Fox Keller argues, is that it is not rooted in the scientific reality of genetic variation. Genetics threatens to be the most sinister example of the power of a discourse to produce its normalized subjects – both through its representations of what it is to have 'normal' genes and through its power to control the materialization of genetic combinations through selective abortion or IVF screening. The Human Genome Project claims to map 'what it means to be human,' to create a 'base-line norm' for humanity, even though there are as many as three million base pairs of difference between individuals (Keller 294). As the examples above suggest, the decisions regarding which of these variations are 'normal' and which 'diseased' will be made based on social classifications, which vary with time and space. The power of genetic manipulations could provide us with the tools to ensure that the genetic ingredients for our current social configurations are those which are passed on to future generations.[21] A further conundrum, as Keller goes on to argue, is the issue of where the agency for making choices about normality or abnormality resides, if all humans – who could make such choices – are simply expressions of the determination of their genes. It becomes overwhelmingly important to pay attention to 'how the authority for prescribing the meaning of "normal" is distributed' (Keller 299).[22]

If we want this power to decide distributed beyond scientists, it becomes crucial to ensure that genetic possibilities and consequences are discussed in discourses beyond scientific journals. Lee Silver's book *Remaking Eden* is an example of popular science writing about genetics. He imagines the future polarization of society between the GenRich and naturals and the eventual elimination of naturals as evolution 'naturally' moves on to the superior GenRich as the next evolutionary step.[23] Further, he imagines humans of the future having abilities such as seeing in ultraviolet and infrared light ranges or hearing radio waves directly through their ears, abilities that he believes can be created through the combination of human and animal genes. I cite this example to argue that in attempting to understand the social effects of genetic modification, non-fiction writers are grappling with the questions in the mode of science fiction. I believe that this blurring of genre boundaries is significant because, in terms of the discourse on genetics, it is becoming increasingly difficult for the non-scientific reader to distinguish between 'real' risks and 'purely fictional' ones. This blurring of boundaries can work in the reverse: while readers may get the 'wrong' ideas about science from reading fictional accounts, they may also get the 'right' ideas about the social implications of science by reading narratives which explicitly engage with the ethical dimensions of technology, something many science fiction narratives do. Butler's *Xenogenesis* novels, while not explicitly concerned with warning us about the medical risks of new genetic technologies, are concerned with warning us about the social and cultural risks of a culture in which genetic information is the 'truth' about individuals. Through displacing the genetic science to an alien technology, Butler encourages her readers to see ourselves as the objects acted upon by genetic technology rather than the subjects who choose how to use it.

What is clear from Butler's narrative is that the assumptions we bring about the body and its meaning will inform the choices we make as we reshape our social world with genetic technology. Grosz has argued that feminists need to be aware of assumptions about the body and identity that ground the Western philosophical tradition, because 'insofar as feminist theory uncritically takes over these common assumptions, it participates in the social devaluing of the body that goes hand in hand with the oppression of women' (*Volatile* 10). Similarly, Butler's narrative shows us that we must be aware of the assumptions about the body that are drawn upon by genetic essentialism if we wish to shape our engagement with genetic technologies on other premises.

The key issue driving the narrative in *Dawn* is Lilith's status vis-à-vis the Oankali and the humans. In short, is she a traitor to her race? This is certainly the way that many of the humans feel about her. The Oankali never mistreat the humans in any way – in fact, they saved them and their planet from the consequences of a human-caused nuclear war – yet many humans fear and resent the Oankali.[24] The main reason for this is the threat that the Oankali present to human identity: humans will continue, but their offspring will be genetically different because of mixing with Oankali genes. The hatred directed toward Lilith is similar to the homophobia analysed by Patrick Hopkins in both its expression and its source. In his essay 'Gender Treachery: Homophobia, Masculinity and Threatened Identities,' Hopkins explores the complex intersection of identity, gender, and sexuality. He argues that 'Because personal identity (and all its concomitant social, political, religious, psychological, biological and economic relations) is so heavily gendered, any threat to sex/gender categories is derivatively (though primarily nonconsciously) interpreted as a threat to personal identity – a threat to what it means to *be* and especially what it means to *be me*' (171). It is this threat to self, Hopkins suggests, which explains the violence that so often accompanies homophobia. Similarly, the Oankali threat to change human morphology and genetic identity is seen by many human characters to be a threat to the continued existence of humans. Lilith's changed body and hybrid offspring are read as acts of treachery by those humans who insist on some sort of purity of categories. Her very existence beyond the categories challenges the security of those who reside within them.

The Oankali threaten what it means to be human and most humans, including Lilith initially, never pause to consider whether this might be a good thing. Lilith is labelled a traitor because she appears to participate in activities that threaten the constructed boundary between human and non-human, much as the gender traitor threatens the boundary between male and female. Nikanj, Lilith's ooloi mate, tells her, 'Our children will be better than either of us ... We will moderate your hierarchical problems and you will lessen our physical limitations' (*Dawn* 243); Lilith, however, to a large extent shares the Resister perception that the Oankali are 'going to extinguish [humans] as a species by tampering genetically with our children' (*Dawn* 227). The Oankali attitude toward humans brings us back to the discourse of genetics and its eu-genetic arguments about improving the species. The Oankali, with their ability to read genes, are a biological equivalent of genetic screening. They believe that human genes doom humanity to hierarchical competitiveness and self-destruction. Significantly, however, what the Oankali most value about human genetics is its ability to produce cancer, a genetic trait that under Oankali control can become a tool of regeneration rather than a lethal growth. As Cathy Peppers points out in 'Dialogic Origins and Alien Identities in Butler's *Xenogenesis*,' this new attitude toward cancer is representative of the novels' overall focus on new origin stories and deconstructing boundaries between subjects. Peppers argues, 'Cancer cells are not wholly other, but exist precisely on the border of me/not me. In revaluing cancer, the text is also therefore valuing "mutation" and "boundary crossing" identity' (53). This revaluing of cancer is part of a larger concern with choosing complexity and connection over hierarchy and binary division that lies at the novels' core. This need to move beyond binaries applies as well to the Oankali perception of the human flaw.

The novels are full of examples that prove the Oankali correct in their assessment that humans are doomed by their genetic inheritance. The first human that Lilith encounters following her isolation among the Oankali tries to rape her. The first group of humans that Lilith awakens and trains soon form competing cliques between those who believe Lilith's 'story' of aliens and those who believe she is lying to them. Curt, the male leader of the group, who resists her most actively, insists on mandatory heterosexual pairing among those awakened. Leading an attempted rape of a woman who refuses sexual advances, he claims 'We pair off! ... One man, one woman. Nobody has the right to hold out. It just causes trouble' (*Dawn* 173). Members of this first group of colonists kill Lilith's human lover, Joseph, because his Oankali-given improved healing abilities make them fear his difference. Once the humans have returned to earth from stasis on the Oankali ship and leave the Oankali to form Resister settlements, these settlements begin to raid and vandalize Oankali-human settlements and, later, each other. Guns are among the first technologies that are reinvented by the humans. Roaming groups of Resister men buy or steal women from the permanent settlements, and steal children from the Oankali-human settlements to sell them to the infertile humans. A human who steals Lilith's son Akin later shoots an animal Akin is petting, simply for the pleasures of killing it and of frightening Akin.

Hoda Zaki also discusses the importance of genetics as a discourse when reading this trilogy. She argues that the novels accept the genetic essentialism premise of the Oankali, both that humans are inherently violent and that men are more violent than women (241). She believes the novels narrate the human 'incapacity to change in response to radically altered conditions' (242). As I outlined above, many critical responses to the novel that focus on the cultural rather than biological characteristics of Butler's humans argue in direct opposition to this position. Peppers, in particular, reads the novels' new origin story as an argument that the return to earth can be a return to something other than the Stone Age; it doesn't have to be about male violence and female reproductive captivity ('Dialogic Origins' 58). I believe that the novels also suggest a much less deterministic attitude toward human biological capacity than Zaki's argument admits. The Oankali may be genetic essentialists, but Butler's readers are encouraged not to be.

Despite the many examples of humans reverting to violence and hierarchy, Butler nonetheless retains sympathy for them and devotes most of the second novel in the trilogy to arguing the case that they deserve a second chance. The Oankali have reserved one-third of their number, a family division called the Akjai, to continue the genetic mix of Oankali as they existed prior to human contact. This is a type of insurance policy, in case the genetic combination that emerges from breeding with humans proves to be unstable and becomes extinct. Akin believes that 'There should be a Human Akjai! There should be Humans who don't change or die – Humans to go on if the Dinso and Toaht unions fail' (*Adulthood Rites* 371). Akin doesn't share the genetic essentialist perspective that characterizes most Oankali in their assessment of human potential and he comes to champion the Resister cause:

> Who among the Oankali was speaking for the interests of resister Humans? Who had seriously considered that it might not be enough to let Humans choose either union with the Oankali or sterile lives free of the Oankali? Trade-village Humans said it, but they were so flawed, so genetically contradictory that they were often not listened to. He did not have their flaw. He had been assembled within the body of an ooloi. He was Oankali enough to be listened to by other Oankali and Human enough to know that resister Humans were being treated with cruelty and condescension. (*Adulthood Rites* 396)

As Michelle Green argues in her article 'There Goes the Neighborhood,' Butler makes 'tricky' use of essentialism in her novels (167).

Despite the many examples of humans reverting to predictable form in the novels,[25] she never collapses the narrative to the genetic essentialism perspective of the Oankali. Despite an almost overwhelmingly negative portrayal of the prejudice and violence which characterize the 'pure' humans, Butler refuses to endorse the eugenics perspective that it is a kindness to curtail the reproduction of such 'defectives.'

Butler's trilogy suggests that while genetics may offer us clues to our identity and potential, it is not a script that determines our fate. While the body is important to understanding our identity, social experiences shape and change us as well. Akin gains his pivotal role as spokesperson for the Resisters not through sharing genetic material with humans – as do many other Oankali-human children born before him – but through his experience of human culture during the time he lives in a Resister settlement after being kidnapped. By casting the genetic engineers as aliens who threaten our continued existence as a species, Butler's discourse is able to focus our attention on the risks of genetic manipulation. Instead of representing a flawed 'them' – the disabled, homosexuals, etc. – marked as other and proposed for elimination from the gene pool through selective abortion or genetic manipulation, Butler suggests that the entire human genome is flawed. This representation brings home the issue at stake in genetic discourses that advocate the purging of this or that gene from the gene pool: the elimination of people like me. In engaging with the world of Butler's novel, no human is able to consider his or her self as exempt from this 'me,' as the privileged possessor of 'normal' genes. *Xenogenesis* helps everyone empathize with the perspective of the so-called genetically flawed.

Even after the Oankali have approved Akin's plan for the Mars colony, they still believe that it is a bad idea, doomed to failure. They argue that they '*know to the bone*' that it is a mistake to restore fertility to unchanged humans, for the human race inevitably will destroy itself again. To the Oankali, allowing unchanged humans to breed is 'like deliberately causing the conception of a child who is so defective that it must die in infancy' (*Imago* 518). However, despite this certainty, the Oankali respect Akin's right to speak for the humans based on his experience of living with them. The reference to the humans as an infant recalls the use of selective abortion in contemporary genetic practices, and encourages the reader to consider the practice in light of this ethic. The non-disabled do not have the right to decide whether or not disabled lives are worth living. Further, the conviction that the Oankali have in the correctness of their assessment – they know through their innate ability to read the genes – parallels the common perception of the adequacy of genetic assessments, a belief that is not defensible.[26] Through the voice of Akin, the trilogy challenges this assessment, arguing, 'Chance exists. Mutation. Unexpected effects of the new environment. Things no one has thought of. The Oankali can make mistakes' (*Adulthood Rites* 488). Even if we choose to believe that it all comes down to our genes, we are cautioned against believing that everything can be known in advance. Green reads the trilogy as a demonstration of 'how human agency can triumph over prejudice, violence and essentialism' (187) and cautions against reading it as a serious discussion of human genetic flaws. In my view, the novel is a serious discussion of the flaws of the discourse of genetics and its presumptuous separations of the flawed from the normal.

In addition to its representations of genetics and suggestions of human genetic flaws, *Xenogenesis* deals with questions of the body through the representation of Lilith's body and the question of Lilith's agency in her first pregnancy with the Oankali. As well as being able to read the human genetic code, the Oankali believe they have some skill in reading the desires of the body. Although she no longer has control over her own fertility, Lilith has been promised by her ooloi mate, Nikanj, that she will not be made pregnant until she is ready, and that she will be the one to decide her readiness. However, at the end of the first novel, Nikanj tells Lilith that it has made her pregnant. Lilith is upset about being forced into a pregnancy against her will, but Nikanj explains that it was only acting on Lilith's silently expressed desires: 'you are ready to be her mother. You could never have said so ... Nothing about you but your words reject this child' (*Dawn* 242–3). Later, another human asks Lilith if it were true that she wanted to become pregnant at that time. Lilith responds, 'Oh, yes. But if I had the strength not to ask, it should have had the strength to let me alone' (*Adulthood Rites* 269). Lilith's dilemma raises the question of relationship between body and subjectivity. Lilith's body expressed her desire to have a child, but it did not express the full extent of her subjectivity – her simultaneous and contrary desire to resist interbreeding with the Oankali. The Oankali romanticize the body, believing that it inevitably speaks the truth while words and consciously expressed desires can be used to deceive, even to deceive oneself. This representation of the body both reinforces the notion of a mind/body split, and recalls various discourses in which the 'truth' revealed by the body is used to undermine the agency of the subject. Nikanj discounts Lilith's statements – 'nothing about you but your words reject this child' – and in doing do it erases Lilith's agency, at least as far as she is concerned.

Human concepts of intersubjectivity and agency rely on representation and language to fill the space between one person and another; we have no concept that allows for an intercorporeal sense of communication. If Lilith's body speaks a truth, the question becomes, whose truth does it speak? As a woman and as a black person, Lilith already has had the experience of having her body positioned in the discourse of another. Boulter notes that Lilith's ambivalent response to pregnancy can be linked to the history of slavery and the conditions under which slave women had their children. In this context, the slave-owner has control of the woman's fertility, a control expressed by forced breeding, rape, the status of children as increased wealth for the slave-holder, and the possibility that one's children might be sold away (177). Susan Bordo describes the risks to women in a context in which their bodies are viewed as speaking more truthfully than their self-representations (76), and Franz Fanon explains the consequences for black subjectivity that emerge from being socially positioned in terms of one's body first and one's individuality second. In both cases, a culturally informed 'reading' of the body's meanings is taken to be more relevant than the self-representations of the subject in the body. Just as Butler cautions us against accepting the Oankali's genetic essentialism, she also warns us against romanticizing the body as a pure expression of nature and truth. The issue of the truth that Lilith's body speaks suggests, as does Jacqueline Rose, that we need to develop a more complex understanding of subjectivity and agency in which 'desiring something is not the same as, and is not reducible to, wanting it to happen' (235). We are again returned to Grosz's image of the Möbius strip and the necessity to understand the body as both a product of culture and a natural material, as both the self and matter shaped by self.

Within the context of the trilogy the Oankali are shown to make mistakes in their interpretation of humans. They believe that they will protect Joseph, Lilith's human lover, by increasing his strength, but instead he becomes a target for those who fear difference and change. They underestimate the vehemence with which male humans respond to Oankali sexual couplings that cause the males to feel 'taken like a woman' (*Dawn* 201) and tainted by homosexual desire (as the couplings also involve a male Oankali). These errors in interpretation warn the reader against accepting the Oankali reading of humans through their bodies, and return us to the discourses of genetics and attempts to read human potential or fate through genetic predispositions.

The method of reproduction in Oankali culture – through the selection of genes under ooloi control – suggests analogies with ARTs (assistive reproductive technologies). The five parents of Oankali reproduction suggest the possibility of separating social from biological parenting, a possibility also offered by ARTs. Marleen Barr argues, 'Butler's alien Oankali, who alter humanity's reproductive capacity, are an exaggerated version of real-world biological alterations of women's bodies' (84–5). Citing Gena Corea's critique of ARTs as something that moves reproduction from the control of women to the control of patriarchy and fears that practices such as surrogate pregnancy will produce a 'Brothel of Wombs,' Barr argues that Lilith's enforced pregnancy is a condemnation of current reproductive technologies as they are used in the hands of 'technodocs' (88). Barr's argument becomes somewhat confused, however, as she tries to sustain a reading of the novels as both a critique of ARTs and a representation of the Oankali as positive, life-revering entities whose 'plan is for the best' (85) and who provide 'a potentially positive new space in which to be free from patriarchy' (86). As I have already argued above, I believe that Butler's portrayal of the Oankali is more ambivalent than Barr's assessment suggests. While Butler is clearly critical of the self-destructive tendencies of humans, she sees even greater risks in the hubris of assuming that there is an all-knowing subject position – alien or scientific – that could presume to correct these faults.

Dion Farquhar argues for a more positive reading of the potential of ARTs to restructure social relationships around parenting and reproduction. She suggests that they allow us to divide among separate people the various social functions of three types of maternity – 'genetic/chromosomal, uterine/gestational, and social/legal' – and two types of paternity – 'genetic/chromosomal and social/legal.' Dividing these roles among different individuals thus exposes 'the constructedness of "natural" laissez-faire reproduction of heterosexual intercourse' and allows us to 'enlarge and diversify meanings of kinship beyond the limits of "blood"' (190). One of the key points that Farquhar makes in her analysis of discourses about ARTs is that both those who are in favour of such technologies and those who oppose them rely on a fixed understanding of certain terms: mother as female and housewife, father as male and breadwinner, parent as heterosexual, and family as biogenetic (11). Those feminists who fear the impact that ARTs may have on the social status of women envision the womb as yet another part of women's bodies that can be exploited by patriarchal culture, while other feminists feel that the prospect of ectogenesis may at last free women from the trap of domesticity represented by pregnancy and nursing. In both cases, however, reproduction is still seen as something that primarily concerns women, and both discourses structure reproduction as something that was previously 'natural' and which now moves into the domain of technology and culture. As Farquhar argues, the domain of human reproduction is already something that is cultural rather than natural, and the options offered by ARTs merely make this fact more apparent. The economic contract to reproduce a male's genetic material through a surrogate pregnancy merely makes more apparent the way in which women have been providing this 'service' to men through the contract of marriage. Surrogate pregnancies also break the link between gestational motherhood and social motherhood, suggesting an opportunity to restructure parenting relationships outside of the heterosexual family.

In *Xenogenesis*, Butler's representation of Oankali reproduction and parenting relationships provides an opportunity to explore some of these more positive aspects of ARTs, although Butler's representations themselves remain rather conservative. For example, the genetic mix of five parents in Oankali offspring suggests that both reproduction and parenting can be separated from heterosexuality. Although the moment of conception is separated from sexual activity in both IVF and Oankali reproduction, in both examples – as IVF is currently used – there remains a relationship of sexual desire between the parents. However, this is not necessary to the practice of reproduction in either case, and could be used as a ground from which to argue for the separation of sexuality and reproduction, undermining discourses which rely on the 'natural' status of sexual reproduction to restrict expressions of sexual desire to those activities which lead to reproduction. However, *Xenogenesis* does remain rather conservative in its representations of both sexuality and reproduction in two ways. First, although the coupling of five partners through the ooloi suggests the possibility for homosexual desire, the novels resist open representation of homosexuality. The five-partner coupling *requires* a previously developed heterosexual couple from each species to join together with the ooloi. Once bonded with an ooloi, no human partners – homosexual or heterosexual – can touch one another directly; but only through the 'filter' of the ooloi. This arrangement – and descriptions of the sexual stimulus that the partners experience from the ooloi during sexual activity – suggests that desire continues to be channelled through normative paths in the couplings. Descriptions of desire in the novels either emphasize the continuing desire of the heterosexual couple for one another (positing the ooloi as simply a conduit for this desire whose presence masks the existence of the other heterosexual couple on the other 'side') or emphasize the pleasures of masturbation (situations when a single partner couples with the ooloi alone).

Anxiety about homosexuality is, in fact, one of the key triggers for the anti-Oankali response on the part of Resister humans. Boulter's reading of the novel links the lack of homosexuality within it to its concern with how humans create and maintain boundaries. Insisting upon the 'natural' heterosexuality of humans, the Resisters use the denial of human homosexuality as one of the characteristics that marks the boundary between 'pure' humans and 'tainted' collaborators. Boulter thus reads the Resister demand for heterosexuality as one of the many false constructions of difference that Resisters rely upon to maintain the fragile concept of pure human identity. Although the Resisters insist that the gap between humans and Oankali is the only important category at work, they must 'nevertheless work to suppress differences within, between, and among humans that might complicate the binary between the self and the alien' (Boulter 175) in order to sustain this sense of polarity. This displacement of differences within humanity onto the figure of the alien/collaborator also suggests why race is not a more explicit concern in the novel. The surviving humans come from a variety of racial and/or ethnic (and ideological) backgrounds, but these classifications, which had led to the nuclear holocaust before the arrival of the Oankali, suddenly become insignificant in the face of the alien difference. Although this fact supports Michaels's reading of the novel as demonstrating that racial difference is irrelevant, the need for the Resisters to actively construct and police the larger difference between human and Oankali suggests that irrelevance is not the crucial measure of efficacy in such discourses.

The second way that the novels are conservative lies in their romanticization of the female body and its natural bond of mothering in their representation of Lilith. Although Lilith feels that cooperation with the Oankali is the most pragmatic course of action under the circumstances, during the first novel she shares the Resister perspective that the Oankali's plan will eliminate humans as a species. By the second two novels, Lilith has had a number of children and has come to realize that a part of her and of human culture does continue in these children. At the same time, however, Lilith continues to feel trapped by this relationship to her children: the option of joining the Mars colony is not available to her because she will not leave her children. This change in Lilith's perspective suggests a construction of the female body as somehow more 'natural' than the male body as Lilith is compelled by her biological bond to her children to remain with the Oankali. Human males, in contrast, are considered to be wanderers who do not bond with their children in the same way.[27]

The final way in which the body is important as a motif in this trilogy is in the representation of the first ooloi children to be born of the human and Oankali genetic partnership. Lilith's child Jodahs is the first to move through the puberty-like transformation from non-sexed child to young ooloi (whose neuter gender is still considered a sex by the Oankali). As a construct[28] ooloi, Jodahs's body has remarkable powers of morphological transformation, a heritage of the cancer 'gene' in humans that so intrigued the Oankali because of its powers to alter cell reproduction and growth. Once it[29] enters a stage of sexual maturity, Jodahs finds that its body responds to the desires of those it desires, and changes its shape to become most pleasing to these others. Eric White, in 'The Erotics of Becoming: *Xenogenesis* and *The Thing*,' has pointed out that the construct ooloi are a representation of the subject as always-in-process. The construct ooloi are literal representations of those forces of subject formation in which the body itself changes in response to the perceived demands of the community of others. Jodahs, and later its sibling Aaor, are the embodiment of Foucault's disciplined subject. White argues that construct ooloi represent Butler's desire to articulate a balance between social stability and essentialism. While it is possible for them to change – they are not wholly determined by their subject position – they need community to avoid dissolution, suggesting that the subject must discipline his body to community standards in order to avoid psychosis.

In the absence of mates to bond with, construct ooloi lose their sense of self, their coherent body image. Without external cues indicating which form to assume, the body fails to retain its integrity: The construct ooloi seemingly have no sense of self other than self-in-relation, and without an external shaping force their bodies begin to break down into simpler forms, moving toward individual cells. Current theories of subject formation tell us that our sense of self is constructed out of our identifications, and our ability to enter into human society is predicated upon recognizing a call of ideology that *is* to our unique body and self. Butler's representation of the construct ooloi makes literal this idea of subject formation: the subject itself, its individual identity, and its body morphology change in response to changing social expectations. Jodahs and Aaor become the perfect mates for those they desire, but they risk dissolution in the absence of mates who provide them with a social role.

There are two ways of reading this representation. One is to view it as a negative representation of the normalizing power of ideology, which controls how our bodies materialize and controls the subjectivities that emerge in conjunction with those bodies. However, a space for agency, or at least for social change, is also implicit in Butler's representation because, as White has noted, the construct ooloi are capable of change. From this perspective, the power of ideology to reproduce itself through producing bodies and subjects is shown to be susceptible to rearticulations and new formations. We may thus read Butler's representation as an example of the power of reverse discourse as de Lauretis has defined it. Jodahs and Aaor change in response to a changing context, forming new identifications and new morphologies. If we read the mates that Jodahs and Aaor encounter as a representation of ideology, and Jodahs and Aaor themselves as representations of the human becoming a subject in response to the calls of ideology, their malleability suggests a space for cultural change. In *Discerning the Subject*, Paul Smith describes this tension between agency and the work of external forces in creating the subject as a hold of the ideological that is alienating and contradictory. Drawing on Kristeva, he argues that literature becomes the site where this alienation is thwarted because the text 'comes to constitute something like a borderline or an interface between the demands of the social and subjective existence' (120). The malleability of the construct ooloi in response to different potential mates suggests something of the malleability of social subjects as they move between cultural texts. Butler's construct ooloi become a literal representation of how the subject and the social are mutually constitutive and suggest that, through the production of cultural texts which offer new types of identifications, we may change the subjects formed through these identifications and so change the social context produced by these subjects.

A second way to read the construct ooloi is in terms of Foucault's ideas about bio-power and its ability to constitute subjects through the normalizing power of dominant body images. Again, we can see this representation as a negative indictment of the destructive power of such normalized body images to destroy the identity of those subjects who fail to inhabit them successfully. Jodahs and Aaor literally have no sense of self – no bodily integrity – unless they can construct themselves in response to the desire of another. Gail Weiss's work on body images suggests a different perspective from which we could view the process of inhabiting many body images. Weiss argues that it is normal, and in fact healthy, for individuals to have multiple body images that they move between in response to the context that the body/subject occupies. Weiss does not believe that such multiple body images suggest that the subject is wholly determined by outside cultural forces or that the subject is unable to develop a coherent sense of self. Instead, these images help us to retain a constant sense of self throughout many changes to our cultural context. In fact, she argues, 'it is the very multiplicity of these body images which guarantees that we cannot invest too heavily in any one of them, and these multiple body images themselves offer points of resistance to the development of too strong an identification with a singularly alienating specular (or even cultural) image.' Thus, rather than showing an absence of self, the fluid construct ooloi can be seen to 'destabilize the hegemony of any particular body image ideal' (*Body Images* 100).

From this perspective, we can read the construct ooloi as an opportunity for agency to emerge in the exchange between subject and context. Weiss cautions us against perceiving culturally offered body images as inherently oppressive. The variety of potential identifications prevents a single hegemonic articulation from entirely determining a subject. The construct ooloi are a troubling representation, precisely because they suggest that outside of the desire of another, outside of the community, one loses identity and dissolves. This image is troubling because it forces us to face the subject – that overdetermined and ever-changing entity, structured by cultural inscription – while it refuses to allow us the solace of belief in a stable self. The construct ooloi can ultimately be read as a positive image, both because they foreground the mutual construction of self and the social, and because their malleability suggests that we are not limited to the current cultural formation of our identities. We can imagine an elsewhere and work to materialize it. The fact that the construct ooloi materialize in response to the desire of the other need not be read negatively; it simply suggests that the identities and social realities we produce are always community products. Articulating an identity outside the space of community is psychosis, and the only way our reverse discourses can have political efficacy is if they compel community belief in the newly articulated.

Butler's trilogy suggests that the body and its genetic code are part of the subject, not just a base material house, but that an understanding of the body alone is not sufficient to understand human consciousness. The body, in Butler's work, is a cultural as well as natural product. *Xenogenesis* cautions us against adopting a perspective of either genetic essentialism – in which we believe that all human potential can be predicted through reading our genes – or the perspective of genetic welfare – in which we believe that we can find genetic solutions to social problems. The construct ooloi are another example of genetics producing the unexpected: neither Jodahs nor Aaor was supposed to develop into the ooloi gender, and their fluid morphology in response to partners is an unanticipated ability. The Oankali represent the promise of enhanced health that genetic science offers to us,[30] but Butler insists that freedom from disease should not be purchased at the cost of genetic variations – people – deemed flawed or abnormal. *Xenogenesis* is not a condemnation of genetic engineering and the benefits it may provide to society. In fact, the portrayal of the Oankali is far more positive than the portrayal of most humans. However, the trilogy does caution against too strong a faith in genetics to unlock the mysteries of humanity and warns of the possibility that genetic information could be used in socially repressive ways. We never get to discover the fate of the Mars colony: perhaps it self-destructs as the Oankali predict, but hope that it will succeed remains.

Butler narrates with sympathy the fears of those humans who have become obsolete as humanity moves on to its next genetic configuration of human-Oankali hybrids. Clearly, the practice of genetic engineering is one that is here to stay in our culture, a Pandora's box that cannot be un-opened. However, this fact does not mean that the discourses of genetic essentialism and genetic welfare must remain hegemonic in our engagement with this technology. Through my reading of Butler's trilogy, I have suggested some of the risks associated with these discourses. Butler shows a space where change could take place, a chance to engage with the possibilities offered by genetic engineering within a context of community and social relationships. The Oankali 'genetic engineers' – the ooloi – are not revered specialists who isolate themselves from the larger community. They engage in their work in a context in which they have relationships with those they treat. Further, the Oankali, despite their conviction of the absolute correctness of their genetic assessments, do not allow this conviction to erase the perspective offered from social rather than 'scientific' assessment.

Although fears of genetic engineering are often expressed in terms of its ability to create grotesque, transgenic biological monsters, the true risks that it represents are the social problems it may contribute to or create. Although the discourse of genetics suggests the nature/nurture debate is over and that nature has won, it is still the social choices we make regarding what is 'natural' that matter. Nurture, or the social, still controls how and where 'natural' bodies can materialize. The new abilities provided by genetics to manipulate nature make it all the more imperative to engage critically with our discursive representations of the 'natural' and the 'normal,' since we now possess the power to remake nature in our image. The necessary supplement to the science discourse of genetics and its representations of the 'natural' body is an understanding of how culture also contributes to the body.

We do not need to fear the monstrous others that might emerge from genetic engineering; instead, we should beware the desire to define a 'base line' of normality for the human genome and limit to the very cellular level those bodies we allow to materialize. Even that most favourite literary example of the monstrous – Frankenstein's creation – was made monstrous by his socialization, not by his biology. Almost in spite of itself, the discourse and practice of genetics is an exemplary representation of the reality that the body is a cultural as well as a natural product. In *Xenogenesis*, the human Resisters are opposed to genetic mixing because they fear that their children will not be human anymore. The implications of the practice of technology exhort us to pay close attention to how we choose to construct this notion of what is 'human' or we, too, run the risk that our children will not be human, because they will lack the diversity of contemporary humanity.

# 3 Iain M. Banks: The Culture-al Body

By disrupting notions of a stable, autonomous, uniquely human self, post-humanist theorists hope to create the conditions for the emergence of less hierarchical and less violent social and political relationships.

Ann Weinstone, *Avatar Bodies*

Iain M. Banks's *Culture* novels depict a far distant civilization of tremendous technological power and social enlightenment. The Culture has eliminated material want, waste, scarcity, the need to work for one's livelihood, and discrimination based on race, gender, sexual orientation, or class. Humanoid citizens of the Culture inhabit radically genetically and surgically modified posthuman bodies that are as indestructible as those of video-game heroes: one can always start over at full strength after being treated for any injury short of brain death. Bodies are altered by a process called genofixing that provides increased longevity, freedom from disease, the ability to switch from male to female morphology, and access to a variety of chemicals that alter mental and emotional states through secretions of drug glands. Days are filled with intellectual and/or hedonistic pursuits as Culture citizens inhabit various technological and architectural marvels (on planets, space stations, and ships) designed by the AI Minds who share membership in the Culture with these radically redesigned humans.

Upon initial assessment, Banks's Culture shares many features with the world offered by Octavia Butler's Oankali; both are utopias in which everyone is free from disease and want. The Culture is characterized by equality – machine sentience is fully recognized as constituting personhood – and there is no material want. A more accurate parallel for the

Culture might be found in TV's *Star Trek* universe. However, Banks's representation does not share *Star Trek*'s non-intervention directive. Good works are a part of the Culture's mission, and they define good works in terms of policing the universe to ensure that other civilizations move in the direction of the enlightened standard they represent. William Hardesty has noted in 'Mercenaries and Special Circumstances' that Banks's Culture novels explore the tensions between surface narratives of benevolent imperialism and counter-narratives that problematize these surface narratives. Although the Culture seems to be represented as a utopia, Banks generally places characters that are outside or critical of the Culture at the centres of the novels and stories set in this universe. In 'The State of the Art,' a novella that considers the possibility of the Culture intervening on earth in 1977, the tensions in Banks's work are succinctly represented by the conflict between the characters of Sma and Lintner. Each believes that earth will be ruined, one by the Culture's interference and the other by its failure to intervene. Lintner argues for leaving earth as it is, 'natural,' acknowledging that this natural state will inevitably include 'murdering and starving and dying and maiming and torturing and lying and so on' (113), while Sma argues for 'maximum interference' (137) with the natural state of things.[1]

The grounds upon which Sma and Lintner base their arguments – the value of the state of nature – are important to contests over the meaning of embodiment and the desirability of the posthuman. One of Lintner's claims is that humans on earth represent a more moral and vibrant way of being precisely because their bodies remain entirely a product of nature, not contaminated by genofixing. He argues that the Culture is representative of 'the self-mutilated, the self-mutated' and that humans 'are real because they live the way they *have* to. We aren't because we live the way we *want* to' (156, emphasis added). These tensions, between the Culture and its others, between a benevolent vision of imperialism and a critical response to it, and between the 'natural' body and the culturally produced one, are all linked. This chapter examines representation of embodiment in the first three Culture novels, *Consider Phlebas* (1987), *The Player of Games* (1988), and *Use of Weapons* (1990). I will situate my reading of the first three Culture novels in the context of Banks's outline of the Culture's social and political structure in his essay 'Some Notes on the Culture.'[2] Reading the tensions at work in Banks's novels suggests a larger political context for struggles about the category of the posthuman.

Banks represents the Culture as 'an expression of the idea that the nature of space itself determines the type of civilisations which will thrive there' ('Notes'). The political organization of the Culture and its absence of conflict are rooted, Banks argues, in the nature of space-faring civilizations that are based on mobile habitations. The very mobility of these habitations, he suggests, prevents any territorial competition between them. At the same time, the hostile environment of space and the need for cooperation among those residing on a space station or ship to guarantee the survival of everyone on it ensures social stability within the ship or station itself. Thus, the model of social organization upon which the Culture is based is very similar to the model of the social contract, cooperation for the sake of ensuring one's own survival, which informs liberalism. One of the key utopian aspects of the Culture is that 'nothing in the Culture is compulsory' ('Notes'). Banks believes that social ostracism will be a sufficient threat to make certain that the over-riding concern of everyone will be to preserve social stability. In Banks's utopia, the absence of material deprivation has eliminated the motive for crime.

What Banks does not draw attention to in his non-fiction writing on the Cutlure is that this plan for social stability is dependent upon a kind of social homogeneity. In his 'Notes,' he stresses that diversity and the freedom of each individual to pursue his, her, or its own path are the cornerstones of the Culture. The Culture is typically represented as a product of the sentient Minds whose intelligence exceeds that of any other organism, and Banks claims that 'a universe where everything is already understood perfectly and where uniformity has replaced diversity, would, I'd contend, be anathema to any self-respecting AI' ('Notes'). Despite this claim, however, the picture of the Culture that emerges from the novels is one of a strange sort of conformity within diversity: Culture citizens are free to explore a wide variety of sexual combinations, lifestyle choices, and hobbies, but certain moral and ethical principles are entrenched.

Banks's Culture is in fact the progeny of liberal humanism. Like liberal humanists, the Culture believes in the infinite perfectibility of humans and human culture through the exercise of reason.[3] The Culture emphasizes freedom of choice and the absolute right of ownership of oneself as the core to citizenship. Although the Culture is not strictly humanist in the sense that it recognizes sentience other than human (most importantly the Minds), it models its inclusion of Minds within citizenship on a version of self that emerges from humanist discourse, most specifically the ideal that human being is defined by the ability to

think and reason. The Culture emphasizes the value of scientific knowledge, of experimentation and investigation of the world, and sees technological progress as inherently good, working to improve the lives of its
citizens. The safety of Culture citizens because they are always in possible telecommunications with Minds (and thus cannot be lost, or left
injured and alone, etc.) is emphasized at many points in the novel, often
in stories of minor characters which are not taken up at length in the
main narrative. Technology is an integral part of the lives of Culture citizens, embedded even within their bodies, and it is always presented as
something that contributes to their comfort and happiness. Like the
humanist tradition, the Culture measures value and happiness in terms
of the well-being of individuals rather than in collective terms.

The model of political inclusivity and of centrality of rights and individual freedom in the concept of citizenship shows the Culture's roots
in liberalism. The Culture is based on a model of tolerance and freedom for almost any lifestyle choice, up to the limit set by John Stuart
Mill that the exercise of one's freedom should not impinge upon the
freedom of others. The Culture also appears to share Mill's supposition
that if individuals are left alone as much as possible to exercise their
own talents and creativity, this will lead to the moral and intellectual
progress of the society as a whole. The model of the autonomous individual who owes nothing to society for his or her capacities but who
equally is owed nothing by society for his or her well-being is central to
most versions of liberalism, beginning with Locke and continuing to the
neo-liberalism informing American politics today. The combination of a
liberalist ideology with the non-scarcity society of the Culture works to
conceal one of the chief problems with liberalism, that it replaced a
social system that emphasized the community and a series of mutual
obligations which bound people together with one that emphasized the
autonomous individual pursuing only his (the gender-specificity being
pertinent for early liberalism) self-interest. The fate of those unable to
compete in a marketplace society is simply not addressed by the novels,
since the Culture is a post-scarcity and post-money society. Part of what I
am trying to draw attention to in my critique of the Culture and its links
to liberal humanism is how the discourse of liberalism enacts a similar
erasure, unable to acknowledge the ideology of capitalism from which it
grows and the injustices and contradictions of this system. The construction of the post-scarcity society is thus a symptom of what neither the
Culture nor liberal humanism can acknowledge.

The fact that the Culture novels focus on characters peripheral rather than central to the Culture suggests that Banks's work might be read both as a critique of some of these limitations of liberal humanism and as a recognition of the appeal of the ideal. Although absolute freedom to do as one chooses (without harming others) and absolute freedom from material want should in theory lead to the pinnacle of human fulfilment and achievement, we find instead that most Culture citizens lead quite dull lives, and those who are interesting enough to become the focus of the novels are those who flee the comforts of this sometimes too bland utopia.

The Culture and liberal humanism share an abstract and idealist concept of the human, and this is part of what contributes to conformity within diversity. When humanism began to emerge in the Renaissance, what was emphasized was the common 'essence' of mankind, a shared set of capacities and tendencies that qualified one as part of the human species. Similarly, when the free individual with a right to his body and the fruits of his own labour was theorized in the seventeenth century, this model, too, was of an abstract 'human' without reference to gender, race, class, or other contingent specificities. Of course, both humanism and liberalism have developed over time and this development has included paying due attention to problems such as the fact that humans were defined by their reason in an era when women were also defined as inherently less capable of reason, or the fact that the foundations of citizenship in self-ownership and individual autonomy were articulated by slave-owners who didn't perceive any contradiction in these circumstances. Similarly, political franchise in Western nations no longer requires property qualifications, but the idea that a person could not truly be free and exercise his democratic choice unless he had economic autonomy was long a premise of liberal thought.

Banks's Culture seems to be based on the ideals of liberal humanism: individualism, freedom of choice, and diversity. At the same time, Banks is able to use science fiction tropes to erase many of the disparities that trouble this system of thought. The Culture lacks scarcity and thus we are not forced to confront the fact that our ideas of individual rights emerged first of all from the right to own property, to have what others do not. Similarly, as all socially necessary work is accomplished either by citizens, including Minds, who enjoy such work and choose to do it, or else by non-sentient robots, we need not recognize that the non-property-owning subject is required to alienate a part of himself through selling his labour on the marketplace. The parameters by which sentience is defined are not discussed in the novels, but it is worth recalling that similar boundaries have been used to construct women or non-whites as less than male European subjects when their economic exploitation was 'required.' The fact that technology can provide solutions to all humans needs means that the Culture is never forced to confront the central inequity that lies at the heart of liberal theory, which is how it is used to justify economic disparity. Macpherson explains how this connection between capitalist exploitation and liberal theory is necessary, not contingent. The liberal view that all men are by nature equally rational means that they are all equally capable of accumulating capital through their labour, and thus 'those who have fallen permanently behind in the pursuit of property can be assumed to have only themselves to blame. And only if men are assumed to be equally capable of shifting for themselves can it be thought equitable to put them on their own, and leave them to confront each other in the market without the protections which the old natural law doctrine upheld' (245).

The limitations of liberal humanism emerge when one considers concrete specifics – such as are possible within a novel. The theory itself of autonomy, individual freedom, and the absolute value of the human individual is appealing in the abstract. However, as the exaggeratedly perfect technology and wealth of the Culture suggest, it is a system that works to the benefit of 'most' humans only if rational thought and scientific progress can solve all the material problems of scarcity and distribution of resources. In fact the Culture is so perfect that it seems to have even solved the problem of death and decay. A society based on individual autonomy such as the Culture is possible only given the tremendous technological advances made possible by the Minds. Thus, I take the Culture to be a sort of thought experiment about what a utopia of rational, technological perfection might be like. The tension between the ideals the Culture represents and the focus of the story on its periphery suggests that part of what these novels are doing is critiquing the limitations of this sort of liberal humanist utopia. The main source of this tension is the gap between the ideal and the material, the beauty of the vision of the Culture's values and their impossibility unless one posits a world in which technology can eliminate scarcity. One of the ways this tension is played out in the novels is through how bodies are represented within them. Embodiment brings materiality back to the human subject, and thus looking at how bodies are presented in the novels points to problems with trying to realize the ideals of liberal humanism.

The Culture's view of human subjects is a very rationalist one in which actions are guided by the relative weight of gains and consequences and where absence from material want eliminates crime. Banks's Culture does not admit of a space in which desire itself might produce actions deemed 'anti-social' and in which infinite freedom to pursue diversity might lead to conflicts over ethics. Banks's vision of the Culture is founded on the belief that a common and stable ground for ethics would emerge and be endorsed universally by all free subjects. As in liberal humanism, the individual and his or her freedom take precedence over community, a perspective that is buttressed by a faith that a collection of rational individuals each pursuing his or her own self-interest and personal improvement will 'naturally' lead to a good community. This emphasis on individual fulfilment shows what is appealing about the ideal, but the absence of structural impediments to Culture citizens realizing their personal freedom points to one of the weaknesses of liberal humanism's ideals, its blindness to the way freedom and choice are limited by structures which distribute power and resources unequally.

The tension among the competing beliefs in individual freedom of expression, in the possibility of human perfection, and in stable and universal moral guidelines emerges in Banks's novels through the interaction of the Culture with other civilizations. The most problematic aspect of the Culture is its belief that its moral perspective is so enlightened as to entitle it to interfere in the domestic affairs of other civilizations in order to 'encourage' these civilizations to move toward an ethic that more closely matches that of the Culture. As noted above, the struggle to reconcile this perspective with the liberal humanist belief in individual freedom, a struggle that is also part of the 'real world' history of liberalism and its connections to colonialism and the suppression of other cultures, dominates these works. In his 'Notes,' Banks argues that 'the Culture doesn't actively encourage immigration; it looks too much like a disguised form of colonialism. Contact's preferred methods are intended to help other civilisations develop their own potential as a whole, and are designed to neither leech away their best and brightest, nor turn such civilizations into miniature version of the Culture.' However, the perspective that emerges from my reading of the novels does not support this contention; instead, what emerges is a representation of Banks's Culture citizens struggling with 'white man's burden.'

In 'Culture Theory: Iain M. Banks's "Culture" as Utopia' Simon Guerrier argues that the novels may be understood as critical utopias because they represent both mainstream and dissident responses to the Culture. Critical utopias are those which see 'utopia as "ambiguous" with "faults, inconsistencies, problems, and even denials of the utopian impulse"' which demonstrate 'utopia's "continual deconstruction": the recognition that societies are not static, and need continual reappraisal' (36). It is possible to go even further and read the tensions in Banks's work as indicative of the ultimate failure of any utopia and a warning of the potential dangers of believing oneself to be in one, that is, as a rejection of critical as well as unreflexive utopianism.[4] The seductive danger of any utopia is the desire to convert others and the risk of imperialism that accompanies this evangelical impulse. The central tension in Banks's work between benevolent imperialism and its inevitable discontents is linked to this rejection of utopianism.[5] Banks's works show an attraction to the ideals that liberal humanism represents, particularly the ways in which the doctrine of universal human rights has been used in politically progressive ways, both in expanding the range of those deemed included in the category 'human' and also through the establishment of international standards for the respect for certain basic political freedoms (from torture, of religion, from unwarranted detention, etc.). At the same time, the novels also show the struggle with imperialist tendencies that we must confront in any attempt to build utopia.

The Culture embodies this danger in its imperial project to patrol the universe. The Culture strives to be a utopia, offering consolation, rather than the more challenging heterotopia that disturbs given orders of things.[6] Unlike Octavia Butler's heterogeneous utopia, Banks's Culture is one in which conflict is eliminated by the elimination of cultural (that is, ideological, ethical) difference. At the same time, however, the novels also demonstrate an awareness of this danger, mainly by giving voice to the perspective of those outside the Culture. As a character from a society in which the Culture is attempting to intervene remarks:

> [H]as it ever occurred to you that in all these things the Culture may not be as disinterested as you imagine, and it claims? ... They want other people to be like them, Cheradenine. They don't terraform, so they don't want others to either. There are arguments for it as well, you know; increasing species diversity often seems more important to people than preserving a wilderness, even without the provision of extra living space. The Culture believes profoundly in machine sentience, so it thinks everybody ought to, but I think it also believes every civilisation should be run by its machines. Fewer people want that. The issue of cross-species tolerance is, I'll grant, of a different nature, but even there the Culture can sometimes appear to be

insistent that deliberate inter-mixing is not just permissible but desirable, almost a duty. (*Use of Weapons* 254)

The Culture is a demonstration of the power of culture itself to colonize – to mould subjectivities and hence to create a social order in the image of its ideology. Banks refuses to clearly endorse either the Culture or its critics in his work. Although for the most part the Culture represents values that I would argue are more congenial to most readers, at the same time no adequate response to challenges like the one above is ever provided.

One of the ways that the Culture ensures harmony and homogeneity throughout its broad realm of influence is through the modification of the bodies of its citizens. The human body becomes a polymorphous playground through genetic augmentation: 'machines could do everything else much better than [humans] could; no sense in breeding super-humans for strength or intelligence, when their drones and Minds were so much more matter- and energy-efficient at both. But pleasure ... well, that was a different matter. What else was the human form good for?' (*Use of Weapons* 273, ellipsis in original). Banks indicates that 'virtually everyone in the Culture carries the results of genetic manipulation in every cell of their body; it is arguably the most reliable signifier of Culture status' ('Notes'). This genofixing includes the ability to change the physical morphology of the body, eliminating bodily distinctions as a ground for social discrimination.

It is difficult to be too critical of the Culture, as it does, indeed, seem like an ideal place to live: pleasure is the highest ideal and desire is the only reason to engage in work yet somehow all socially necessary work gets done. However, the risk remains that this perfection can only be achieved through the elimination of difference, and through their focus on the discomfort of characters for whom the Culture's norms are not taken for granted, the novels demonstrate a dis-ease with such homogenous perfection. The male and female distinction remains, but any individual can move from one subject position to another; racial variances are effaced through interbreeding and through the ability to modify one's appearance to any racial characteristics one chooses. Although characters within Banks's novels draw attention to the risks of a cultural imperialism represented by the Culture (as its very name suggests), what is less emphasized in the novels is the degree to which the Culture also acts on human bodies in imperialist ways and with similarly disturbing consequences. In the Culture, it is possible to solve social problems through biology alone. This emphasis on the individual – rather than on structures – is part of what I'm suggesting is a limitation of the Culture that is connected to its liberal humanist tendencies.

Judith Butler has pointed out, in *Bodies That Matter* and in *Contingency, Hegemony, Universality*, that there is a danger in believing that one can arrive at the perfect image of inclusive identity, beyond all repudiations and differences. For Butler, the process of contesting the current normalizing ideal can never end, because 'the ideal of transforming all excluded identifications into inclusive features – of appropriating all difference into unity – would mark the return to a Hegelian synthesis which has no exterior and that, in appropriating all difference as exemplary features of itself, becomes a figure for imperialism, a figure that installs itself by way of a romantic, insidious, and all-consuming humanism' (*Bodies That Matter* 116). This is precisely what has happened in the Culture, whose very mode of embracing contingency and multiplicity and change has ironically become an embrace of homogeneity. It is true that one can be any gender, race, or anything else marked by embodied difference within the Culture, without the experience of discrimination. However, it is equally true that one can be so because such embodied difference is deemed not to matter, because difference has become unity. When one changes body morphology by genofixing, nothing other than the shape of the body is deemed to have changed. Thus, the concrete specificity of living in differently raced or gendered or sexed bodies is effectively erased from the Culture's vision of inclusivity. The body itself becomes the symbol of the Culture's overarching imperialism. All difference is erased as species become members of the Culture, a joining that is marked by genofixing, which makes the individual over into a Culture citizen. While the narratives themselves are critical of the possibilities of imperialism represented in the Culture's 'foreign policy,' they seem less aware of the dangers of imperialism lurking beneath the Culture's conception of embodiment.

One reason for this blindness is that Banks has theorized the Culture based on a Cartesian view that the body is not essential to subjectivity. In his 'Notes,' he writes that Minds 'bear the same relation to the fabric of the ship as a human brain does to the human body; the Mind is the important bit, and the rest is a life-support and transport system.' The body, then, in Banks's explicit statements, is little more than a container for subjectivity: changing it should have no significant impact on identity. However, just as the novels suggest a simultaneous attraction to the ideals of liberal humanism and an awareness of their limitations, the representation of the body within them also suggests an attachment to the notion of a natural body, despite a concurrent desire to equate subjectivity with mind alone. When Sma and Lintner argue about the value of preserving the natural earth in 'The State of the Art,' Sma observes that 'greed and hate and jealousy and paranoia and unthinking religious awe and fear of God and hating anybody who's another color or thinks different is *natural*' (157). In each of the novels, the body bears the weight of expressing these and other limitations of the natural.

Despite the claim that the body is not the 'important bit' of the person, each of the three Culture novels I will discuss reveals an attachment to the notion of body as self, though not necessarily as the 'best' part of the self. Even within the 'Notes,' this tendency is apparent. Banks describes the Culture's sex-changing process in the following terms: 'An elaborate thought-code, self-administered in a trance-like state (or simply a consistent *desire, even if not conscious*) will lead, over the course of about a year, to what amounts to a viral change from one sex into the other' ('Notes,' emphasis added). In this description, we see again the notion of the body as natural, rather than as cultural, and as something that is therefore closer to some pure notion of the Truth. Just as Nikanj argues that the desire of Lilith's body superseded her consciously expressed desires, in the Culture one's unconscious desires will emerge as morphological symptoms of the body as it moves between sexes. Banks also adds, 'usually, a mother will avoid changing sex during the first few years of a child's life. (Though, of course, if you want to confuse your child …)' ('Notes,' ellipsis in original). This notion that the female sex is somehow essentially appropriate to motherhood – even within a social context of continually changing morphologies – again suggests an unacknowledged attachment to understanding our contemporary constructions of the body as 'natural.'

Mind/body dualism has historically allowed some subjects – male, white, heterosexual – to construct themselves as unmarked by the body, while other subjects – women, non-whites, gays and lesbians – are seen as having a closer connection to the body, often expressed as being *reduced* to the body. What this reduction entails is that embodied subjects (those whose bodies mark them as different) are not able to attain true subject status, since subjectivity has been equated with the mind alone. Thus, liberal humanism intersects with the Cartesian mind/body split in a very pernicious way. Liberal humanism emphasizes individual freedom and, like Cartesian dualism, it presumes a neutral and transcendent individual subject who can go about pursuing this freedom, a subject who is able to 'own' himself, which means that his self must somehow be separate from his body. At various times, being 'human' was thus impossible for women (owned by others in marriage), for non-whites (owned by others in slavery), and for the working class (deemed incapable of freely exercising political franchise because they were dependent upon selling their labour to those who employed them and thus didn't truly own themselves freely). So long as the free individual is equivalent to the unmarked, non-embodied mind, some subjects can never attain the status of 'individual' to pursue their freedom of expression and make their choices part of the community of values.

Liberal humanism severs the subject from his or her embeddedness in material circumstances, just as mind/body dualism severs the mind from its relation to the body. The politics of each suggest that the individual has a constant essence or identity regardless of circumstances, and that there is something universal about this essence of being human. Thus, such a politics advocates both individual freedom and individual responsibility, and the absence of government constraints and controls. Like the discourse of genetic welfare, this ideology suggests that social problems can be solved through individual change, effacing the effects of systemic discriminations. This emphasis on freedom and autonomy does not allow for a critique of the structures which limit individual opportunity and choice. What is dangerous is the suggestion that there is an abstract essence of 'human' that can be separated from material circumstances. We cannot refuse to acknowledge the consequences of the fact that this 'essence of human' has historically been constructed from the perspective of white, male, heterosexual, property-owning Western subjects.

The Culture occupies the privileged pole traditionally coded white/male/mind, reducing its cultural others to the body and to a status of less than full subject, thereby justifying the Culture's interventions on behalf of these less than mature civilizations. As I will show in my readings of the novels, the gap between the abstract ideals of liberal humanism and a concrete notion of how to put such ideals into practice creates a tension in Banks's work, one that points to both the reasons why liberal humanism is attractive and also the reasons why it is ultimately inadequate.

In the 'Notes,' Banks is strongly supportive of the Culture's social and ethical system. In each of the novels, however, he represents the viewpoints of characters highly critical of the Culture. *Consider Phlebas* is the novel that presents this viewpoint most aggressively through the character of Horza. Horza is a Changer, a humanoid species with shape-shifting ability, who has decided to join with a non-humanoid alien race, the Idirans, in their war against the Culture. The Idirans do not offer much to recommend them as a species: their political system is rooted in their theology and the belief that they are chosen superiors in the universe. They put this ideology into practice through a policy of never-ending territorial expansion; those 'inferior' species that they encounter are offered the choice of enslavement or extinction. An Idiran who refuses to believe that the Changer is his ally treats Horza himself with contempt and abuse. All other species are simply soulless animals to the Idirans.[7]

Despite this, however, Horza remains convinced that the Idirans are preferable to the Culture, and his rationale for this choice is rooted in the concept of what is 'natural':

> On a straight head count [of those killed] the Idirans no doubt do come out in front, Perosteck, and I've told them I never did care for some of their methods, or their zeal. I'm all for people being allowed to live their own lives. But now they're up against you lot, and that's what makes the difference to me. Because I'm against you, rather than for them ... I don't care how self-righteous the Culture feels, or how many people the Idirans kill. They're on the side of life – boring, old-fashioned, biological life; smelly, fallible and short-sighted, God knows, but *real* life. You're ruled by your machines. You're an evolutionary dead end ... The worst thing that could happen to the galaxy would be if the Culture wins this war. (*Consider Phlebas* 26)

Horza's objection to the Culture is integrally linked to his rejection of machine sentience. He believes that its machines run the Culture – which does appear to be true, to a large extent – and that this fact somehow makes the culture of the Culture opposed to biological life. Again, his objections are expressed in terms of valuing what is natural over what is artificial: '[The Culture] could easily grow forever, because it was not governed by natural limitations. Like a rogue cell, a cancer with no "off" switch in its genetic composition, the Culture would go on expanding for as long as it was allowed to. It would not stop of its own accord, so it had to *be* stopped' (*Consider Phlebas* 169–70).

Granted, it is not appropriate to reduce the perspective of this character to the perspective of the author. In fact, although he is the central character of the novel, in many ways Horza is not portrayed as its hero;

despite his claims to value life he is perfectly willing to kill others in order to further his own ends, and his rejection of machine sentience is represented as a type of racism. The novel is deeply pessimistic as well: not only does Horza die by the end, but we are also informed in a 'historical' appendix to the story that the Changer race was wiped out as a species during this war, although the war itself was insignificant: 'A small, short war that rarely extended throughout more than .02% of the galaxy by volume and .01% by stellar population' (*Consider Phlebas* 490). The final irony is that Horza himself is ultimately revealed to be a product of culture, not nature. Although Horza does not learn this, the reader is told that the Changers are a race created to be weapons in a long-forgotten war. By the end of the book, he and his race live on only through his name, which has been taken by the Mind Horza has been pursuing throughout the novel.[8]

This pattern of focusing on characters who are not truly part of the Culture continues: in *The Player of Games*, the central character, while a Culture citizen, is one who is characterized by ennui as he finds the Culture boring; the central character in *Use of Weapons* is a Special Circumstances mercenary, one of the members of a non-Culture civilization who are used by the Culture to perform interventionist tasks that Culture citizens themselves find rather distasteful. I believe that this curious tension between pro- and anti-Culture sentiments in the Culture novels emerges from Banks's recognition that the concept of the natural body has been deployed in politically oppressive ways in societies characterized by hierarchical division. Banks wishes to develop his utopia based on egalitarian terms and, like Octavia Butler, he seems to feel that the human animal is incapable of producing such a utopia on its own.

The desire to characterize the body as purely natural – the unimportant life-support and transport system – connects the Culture's notion of subjectivity to mind/body dualism. The Culture is controlled by the Minds, the ideal of what humans should aspire to be according to this philosophical heritage in which the body is a base, carnal, limiting prison. Without bodies, the Minds are not seduced into the vulgar life of sensation. The infinitely malleable bodies of the Culture are actually a way of transcending the body, of suggesting that the body does not matter, as its matter can be formed to suit the desires of the mind. Despite the fact that the Culture seems to welcome diversity – in bodies – it is rooted in a concept of uniformity – of values or ethics, of human nature grounded in universal reason. To become a Culture citizen, one must be the cogito, must rise above the 'base' desires of the particular body, and embrace the universal rationalism that is the mind; in practice, this has meant rising above what pertains to marked female, non-white, gay and lesbian, and working-class bodies.

Despite the premise that the Culture's genofixed bodies allow its citizens to pursue what seems to be a harmless hedonism, Banks's novels contain an undercurrent which suggests that the body's desires are not as valuable as those of the mind. This motif is most apparent in *The Player of Games*, a novel that recounts the experiences of a Culture citizen, Gurgeh, in the non-Culture civilization of Azad. Azad is a civilization in which bodies matter very much. Azad has an extremely repressive social system that is founded on discriminations based on body morphology. Azad has three genders: the lowest are the females, who are primarily used for sex; the second are the males, who sell their bodies to the state as soldiers; and the third gender, apices, hold all the political and social power. Apices have sexual organs that combine features of both of what we would call male and female.[9] Azad culture is fiercely hierarchical along their sex distinctions, and is characterized by entertainments that emphasize cruelty toward and domination of the 'lower' sexes.[10] Azad also has a hierarchy of racial differences; the dark-skinned Gurgeh is shocked to find himself in such a civilization when he is warned by his bodyguard drone that he risks death if he lets his cloak fall back to reveal his skin colour while in a certain part of the city. The Culture is opposed to the ethos of Azad and the novel concerns its attempts to undermine Azad's political structure.

Banks portrays the Culture's antipathy to Azad as an attitude that applies to the Culture's machines more than to its humans. When Gurgeh arrives in Azad, the Culture's resident ambassador, Za, complains: "'They're all the same those machines; want everything to be like the Culture; peace and love and all that same bland crap. They haven't got the" – Za belched – "the sensuality to appreciate the" – he belched again – "Empire'" (*The Player of Games* 135). This passage again suggests the mind/body dualism that sustains the representations of the body in these texts. Za's body is emphasized as he makes these comments, and it is emphasized in a negative way. The body puts one at risk; it indulges in gross sensuality and cares more for its own gratification than for social justice. This representation of the body as that which compromises the social structure recalls the way in which discourses representing women and non-whites as more closely bound to the body have been used to deny such 'embodied' subjects access to political and social power. Despite the suggestion in Banks's 'Notes' that social utopia can be achieved, in part, through the body modification of genofixing which could thereby eliminate the grounds for body-based discriminations such as racism and sexism, Banks is not a genetic essentialist. Far from being irrelevant, the body is that which threatens his utopia; Banks's utopia is grounded on pure reason, the suppression and denial of the body, and hence can best be represented by the Minds that dominate it.

It is through the shape-shifting character of Horza that the return of the repressed body can most clearly be seen. Horza can alter the physical morphology of his body to imitate other humanoids of any species. Horza's very being, therefore, is a metaphor for the ways in which the body is cultural, able to be moulded and shaped into different forms by different social circumstances. That Horza remains the same person throughout these various shiftings suggests, as does Banks's argument that the body is not the 'important bit' of subjectivity, an adherence to mind/body dualism. However, Horza is ideologically opposed to the Culture, seeing it as something 'artificial' rather than 'natural,' as *for* stagnation rather than *for* life. This opposition is expressed through Horza's contempt for sentient machines, suggesting that he personally believes that the body rather than the mind is the essential component for 'true' subjectivity. In Octavia Butler's trilogy, we saw that a continually changing body image could be read as something positive, something that allows the body/subject to respond to the demands of context without surrendering agency to external determinants. Both the recognition that the body is part of the subject and the recognition that the body is always-already cultural are important prerequisites to successfully inhabiting a multiplicity of body images without falling into a psychosis of fragmentation. Owing to his insistence that the body should be purely natural and that the forces of nature are more moral than the designs of disembodied consciousness, Horza is unable to hold these beliefs. He must believe in a stable body, one that is immune to the influence of culture.

It is ironic, then, that his 'natural' body is one that continually changes its outward morphology to match that of whomever he is currently impersonating. Near the end of the novel, Horza experiences a moment of horror that reveals that he has not been able to maintain a stable identity throughout the changes to his body's external appearance. Just before he falls unconscious from injuries, 'a look of the utmost horror, an expression of such helpless fear and terror [comes over his face]. ... "My name," he moaned, an anguish in his voice even more awful than that on his face. "What's my *name?*"' (*Consider Phlebas*

478). Horza's fear that he has no identity is reflected in the appearance of his unconscious body:

> The face of the man on the stretcher was white as the snow, and as blank. The features were there: eyes, nose, brows, mouth; but they seemed somehow unlinked and disconnected, giving a look of anonymity to a face lacking all character, animation and depth. It was as though all the people, all the characterisations, all the parts the man had played in his life had leaked out of him in his coma and taken their own little share of his real self with them, leaving him empty, wiped clean. (*Consider Phlebas* 481)

This passage suggests that Horza's body has been crucial to the construction of his self – the self-conscious, unified, and autonomous face he presents to the world. In the absence of an external model, his face becomes 'blank' – there is no internal essence to animate his features. His 'real self' was always a construction, a fabrication of coherence that fragments; it becomes 'disconnected' when he is not consciously connected to a specular image. In his insistence upon the 'naturalness' of his body, Horza has invested too heavily in a concept of self as stable and unchanging; when this belief in his own consistency is challenged, he finds himself without other grounds upon which to build his identity. The subject, his ideologically imparted identity, cannot sustain the illusion of coherence and self-sufficiency in the absence of external stimuli. I am not arguing that it is the fluidity of Horza's body *in fact* that creates problems for his identity, but the fact of this fluidity within a set of cultural values that valorizes stability, naturalness, and consistency. If Horza invested in other values, his fluidity might be read as positively as the ooloi's. Thus, the contrast between these two examples emphasizes the importance of cultural milieu in structuring the experience and meaning of a particular sort of embodiment. No bodies, posthuman or otherwise, have meanings in isolation.

One of the ways in which it becomes clear that the body is a risk for the Culture's version of subjectivity is through the changes experienced by Gurgeh in *The Player of Games*. Gurgeh is the Culture's finest gamesplayer, and he travels to the Empire of Azad to participate in its game of Azad, a game that is used by the Empire literally to pick its Emperor. The Special Circumstances branch of the Culture – that branch which is involved in direct intervention in the internal affairs of other cultures in order to 'guide' them appropriately – believes that it is necessary to undermine the game of Azad in order to destroy the political stability of the Empire and hence pave the way for progressive social change. Gurgeh's character is changed by experiences of his body while he is in Azad. During his time there, his body responds to the pleasures Azad offers – pleasures of domination and subordination – that change him from a Culture-like subjectivity, appalled by the cruelties of Azad, to an Azad-like subjectivity, appreciative of the Empire.

Gurgeh's susceptibility to the Empire appears to be rooted in the 'weaknesses' of his body, his lust for the sensations that accompany being on top in a hierarchical structure. Gurgeh had these tendencies before he came to the Empire but he did not overtly express them by repressing others; once his body undergoes the experiences offered by the Empire, he loses many of his inhibitions.[11] Regarding his obsession with game-playing he says:

> I ... *exult* when I win. It's better than love, it's better than sex or any gland-ing; it's the only instant when I feel ... real. ... Me. The rest of the time ... I feel a bit like that little ex–Special Circumstances drone, Mawhrin-Skel; as though I've had some sort of ... birthright taken away from me. (*The Player of Games* 21–2, ellipses in original)

During his time in the Empire, these tendencies are emphasized and extended for Gurgeh. As he becomes more and more absorbed in the game of Azad, he begins 'thinking differently, acting uncharacteristi-cally. He is a different person' (*The Player of Games* 231). The change that Gurgeh experiences, the alteration of his subjectivity in response to the effects of material reality on his body, can be understood in terms of Charles Peirce's concept of habit-change as it has been elaborated upon and used by Teresa de Lauretis. A habit-change is the third level of inter-pretents that link objects and signs in the external world to their effect on the internal world of the experiencing subject. A habit-change is a change in the subject's tendencies toward action, and it results from a history of previous experiences and exertions.[12] The concept of habit-change provides a mechanism for understanding the ways in which the experiences of the body change subjectivity. As Gurgeh plays Azad, his 'brain [is] adapted and adapting to the swirling, switching patterns of that seductive, encompassing, feral set of rules and possibilities' (*The Player of Games* 232) and he becomes a more feral subject in response. The drone accompanying him observes a 'callousness in his play that was new' and believes that 'the man had altered, slipped deeper into the game and the society' (*The Player of Games* 247). Our narrator, later revealed to be the drone himself, asks us to consider the change in Gurgeh as he learns more of Azad the game and Azad the Empire, considering both 'What will he make of such knowledge? More to the point, what will it make of him?' (*The Player of Games* 232).

The body is less pivotal in *Use of Weapons* than it is in the other two novels, but again Banks suggests that the body is cultural as well as natural and that – to humankind's detriment – the body is an essential part of the self. The main character in this novel, Cheradenine Zakalwe, is a mercenary working for the Culture. The novel narrates two stories – the first in standard chronological order and the second in reverse. The first story concerns Zakalwe's present mission, one for which he has been brought out of retirement. The second story, in reverse chronological order, recounts Zakalwe's past missions for the Culture and leads us back to the traumatic event in his past which caused him to leave his own planet and become a Culture mercenary. Throughout the various descriptions of Zakalwe's past adventures, lots of attention is paid to the details of his bodily injuries: the experiences that lead to his wounding, his subjective experience of pain, and the Culture's near-magical ability to restore his body despite repeated injury. At its most extreme, Zakalwe's entire body is replaced after he has been decapitated and only his head is recovered.

Despite this representation of the body as seemingly irrelevant – its parts literally replaced without effect – Zakalwe's body is represented as an essential part of his subjectivity. At the end of the novel we discover that this character is not Cheradenine Zakalwe – a leader of his native planet – but is instead Elethiomel, Cheradenine's stepbrother who led a rebellion against him. The traumatic incident from the past – in which Cheradenine received the body of his sister, Darckense, fashioned into a chair – is revealed as something that this character committed rather than something he was a victim of. The focus on the suffering of 'Zakalwe's' body throughout the novel has been the story of his attempts to achieve redemption through the suffering of his body. His true morphology is found in his internal body image, one that remains consistent throughout the various Culture-al modifications and repairs to the external material. This body image is represented by a physical scar that he continues to 'feel' even after his entire body has been replaced: a scar he received when a bone-chip from Darckense's leg, shattered by a gun wound, pierced his chest. This incident from their childhood continues to haunt him and remind him of the monster he has become in his ability to make anything, even her body, into a weapon to serve his ends.[13] 'Zakalwe's' subjectivity continues to be formed by images of bodies: his own as containing a reminder of Darckense, and his image of her body formed into a chair that fuels his self-hatred.

The body in Banks's novels is thus both the ground for the subject/self – the necessary image of coherence that the self needs – and the threat to a Culture's most deeply held ideals – which are those of the purely rational mind of Descartes. The contrast between the Culture and Azad represented in *The Player of Games* provides a clear representation of the link between the philosophy of a culture and the physical bodies that can materialize within this culture. Social divisions on the basis of the body's sex distinction characterize the Empire. Because the Empire is so committed to this social structure, it needs to demarcate firm and fixed lines between the sexes. The Empire cannot tolerate the fluid sexual continuum of Culture-al bodies; for citizens of Azad, biology must be destiny, or the entire social order is threatened. The Empire is appalled that the Culture regards 'homosexuality, incest, sex-changing, hermaphrodicy and sexual characteristics alteration as just something else people did, like going on a cruise or changing their hairstyle' (*The Player of Games* 225). The Culture can treat the body as a malleable object, but Azad requires a conception of the body as fixed in nature, a sign of predestined worth. Banks makes it clear that it is an ideological distinction – not a technological one – that determines this difference. In a discussion of the absence of sexual 'switching' in Azad, Gurgeh is informed, 'genetechnologically, it's been within their grasp for hundreds of years, but it's forbidden. Illegal' (*The Player of Games* 75).

A social system that is rooted in body-based discriminations must not allow evidence that such distinctions are constructed and arbitrary to emerge. The Empire uses familiar rhetoric of 'natural' differences to justify its oppression of two of its sexes. However, just as Banks suggests that the maintenance of these differences serves the social status quo, he also reveals that their origin is similarly culturally engineered:

> A programme of eugenic manipulation has lowered the average male and female intelligence; selective birth-control sterilisation, area starvation, mass deportation and racially-based taxation systems produced the equivalent of genocide, with the result that almost everybody on the home planet is the same colour and build. (*The Player of Games* 80)

Rather than evoke the new eugenetic-ist discourse of human improvement, Banks takes us back to the 'old eugenics' of elimination of those deemed socially worthless. The Empire clearly evokes the worst of Nazi social policies. In contrasting this society to the Culture, Banks suggests that the Culture's use of genetic manipulation is comparatively innocent – to increase human pleasure, to eliminate disease, to expand the range of human diversity. However, as I discussed above, the Culture shares with the Empire a concept of the body as natural: as 'naturally' revealing innate superiority or inferiority in the case of the Empire, and as the necessarily repressed expression of all that is natural (animal-like and destructive) in humanity in the Culture's deference to its disembodied Minds. This tension in how the body is represented in the novels is thus another aspect of the ambiguities surrounding the Culture in Banks's work.

Although space prevents me from taking up the other two Culture novels (*Excession* and *Look to Windward*) in detail, these novels also repeat some of the patterns that I have discussed. Both present a conflict between the Culture and outsiders who oppose and challenge the values of the Culture, the aggressive and avaricious Affront in *Excession* (1997) and the conservative and oppressively hierarchical Chelgrians in *Look to Windward* (2000). These two latter novels continue the pattern of giving voice to the non-Culture point of view and exploring some of the limitations of the Culture's version of imperialism; in *Look to Windward* the Culture admits that its miscalculation regarding how best to push the Chelgrians toward a more equitable state has caused a bloody civil war. As well, these two novels suggest that self is equated with mind only as each includes characters who lack embodiment but who can nonetheless be brought back fully based on back-up copies of their minds. The most recently published Culture novel, *Look to Windward*, suggests that whatever critique is offered of the Culture, the moral point of view of the novels is ultimately on their side. At the end of the novel, a Chelgrian character discusses his conversion away from his own culture's values of hierarchy and tradition and toward the Culture's embrace of individualism and equality. He tells us that the Culture was able to turn him from the Chelgrian way simply by showing him 'all there was to be shown about my society and theirs and, in the end, [he] preferred theirs' (356). Another non-Culture character, the Homomdan, Kabe, points out that the title Ambassador which the Culture gives to citizens of other civilizations living within the Culture means 'that person represents the Culture to their original civilization, the assumption being that the alien concerned will naturally consider the Culture better than their home and so worthy of promotion within it' (357). It is possible to read this statement as yet another example of the Culture's hubris, a trait frequently commented upon and mocked by non-Culture characters within the novels. However, the fact remains that the narratives themselves repeatedly demonstrate this pattern of conversion, most dramatically in *Look to Windward* itself.

*The Player of Games* asks the reader to reflect upon the social construction of sex/gender categories and their link to political power in our own world. In a direct address to the reader just before Gurgeh's arrival at Azad, the narrator explains the conventions it (the narrator is a drone) will follow in using personal pronouns to describe Azad citizens:

> I have chosen to use the natural and obvious pronouns for male and female, and to represent the intermediates – or apices – with whatever pronominal term best indicates their place in their society, relative to the existing sexual power-balance of yours. In other words, the precise translation depends on whether your own civilization (for let us err on the side of terminological generosity) is male or female dominated. (*The Player of Games*, 99–100)

Marain, the Culture's language, has a single personal pronoun to apply to any subject with the status of personhood, which ranges across the male-to-female sexual spectrum and includes sentient machines. In the English 'translation' of the novel, the masculine personal pronoun is used to describe apices. The invitation to compare sexual power relations on Azad to sexual power relations in our world, in the context of the obvious constructed nature of sexual difference in Azad, encourages the reader to see the sex/gender system as an expression of culture rather than as one of nature. Finally, this invitation to return to the 'real' world can be extended to other aspects of the novel, allowing the reader to reflect also on the advisability of striving for genetically produced utopias and of denying that the body is part of the self.

The body in Banks's Culture, then, critically engages with the concept of the posthuman in two crucial ways. First, Banks shows us one possible configuration of a posthuman society, one in which it seems possible to overcome social ills through the manipulations of bodies. The Culture would have us believe that it is possible to transcend body-based biases and the social injustices that flow from them simply through changing bodies and making bodies changeable. The limitations of Banks's Culture as a utopia points us to limitations of a model of posthumanism that would see the body as separate from the self, a mere tool. Second, the link between representations of embodiment in Banks's work and the tension it depicts between celebration of utopia and critique of its tendencies toward imperialism points us toward another important way to begin thinking about the category of the posthuman. Rather than thinking of the 'post-' in reference to the human body as site of the humanness now superseded, we might instead think of the posthuman as something beyond or after humanism and the violence implicit in its reduction of a plural world to unitary model of 'true' subject. Like Jones's redefinition of self as *the same as* rather than *distinct from* the world and others, this kind of posthumanism is grounded upon a new conception of what it means to be human rather than a new morphology of humanness. Although Banks's Culture shows us changes in morphology only – and is fearful of the body – the dissidents within his utopia can point the reader toward the importance of beginning to think of the posthuman in terms of new categories of identity rather than new appendages.

# 4 Cyberpunk: Return of the Repressed Body

In pop culture, practice comes first; theory follows limping in its tracks. Before the era of labels, cyberpunk was simply 'the Movement' – a loose generational nexus of ambitious young writers ... [who] found unity in their common outlook, common themes, even in certain oddly common symbols, which seemed to crop up in their work with a life of their own. Mirrorshades, for instance ... Mirrorshades prevent the forces of normalcy from realizing that one is crazed and possibly dangerous. They are the symbol of the sun-staring visionary, the biker, the rocker, the policeman, and similar outlaws.

Bruce Sterling, Preface to *Mirrorshades*

Bruce Sterling's dramatic claim for cyberpunk – that it is subversive of the forces of normalcy, crazed and possibly dangerous – has not been supported by many of the critical assessments of the genre. Cyberpunk is a curious phenomenon within the field of science fiction: it has provoked considerable critical debate and discussion both within and beyond 'fandom,' a debate that seemingly has survived the sub-genre itself.[1] A discussion of the body and science fiction must consider the influence of cyberpunk, a genre best known for its rejection of embodiment and embrace of an existence in cyberspace.

Sterling, the 'movement's' most vocal promoter during its inception, has argued for the subversive potential of the genre, its allegiance with the hackers and rockers who challenge the socio-economic status quo. However, critical assessment of the genre by critics such as Darko Suvin,[2] Istvan Csicsery-Ronay, Jr, and Andrew Ross suggests that the subversiveness of cyberpunk lies more in its style than in its substance. Certainly, marketing itself as a distinctive sub-genre has been a successful strategy for cyberpunk if measured in terms of the attention that the genre has received from academics. Part of the reason that cyberpunk has been the focus of so much critical discussion can no doubt be attributed to the wide range of assessments the sub-genre has provoked. Larry McCaffery, following in Sterling's laudatory vein, has argued in 'Introduction: The Desert of the Real' that cyberpunk offers the cognitive maps that Jameson calls for in *Postmodernism*, helping the human subject orient him- or herself to the world of late capitalism. Scott Bukatman contends in *Terminal Identity* that cyberpunk is one of the cultural representations that reinstalls human agency at the site of the terminal, the very site where technology intervened in and deconstructed the human subject's edifice of its own autonomy and unity. Veronica Hollinger maintains in 'Cybernetic Deconstructions' that cyberpunk marks a moment in SF where humanity and its technology now share the narrative foreground, whereas earlier SF tended to put humans in the foreground and technology in the background. Hollinger's reading in particular points to the importance of cyberpunk as a genre exploring the idea of the post-human. Although the literary moment of cyberpunk may have passed, its tropes have never been more germane, as technologies related to computer communications, virtual reality, and human/machine interfaces (such as Steve Mann's wearable computer) suggest that the aesthetic of cyberpunk may become a literal reality. Cyberpunk has received its share of criticism as well as praise. The two most common criticisms of the genre are that it is merely misogynistic, boys'-own-fantasy escapism, and that it offers only individual transcendence of, not social solutions to, the problems it diagnoses.[3]

In order to understand cyberpunk fiction, it is necessary to understand the concept of cyberspace. Gibson's definition of cyberspace in *Neuromancer* has by now become widely repeated and I will not break tradition by failing to quote it. Cyberspace, writes Gibson, is 'the consensual hallucination' into which a cyberspace cowboy's disembodied consciousness is projected. The appeal of cyberspace is linked directly to the repression of the material body in cyberpunk fiction and, increasingly, in real-world 'cyberspace' encounters in virtual reality or in Multiple User Domain virtual environments on the Internet. Vicki Kirby writes, 'cyberspace is ... the space where the perfect body is paradoxically acquired through an annihilation of the flesh' (132). The world of cyberspace is the consummate world of the Cartesian dualist: in cyberspace, one *is* the mind, effortlessly moving beyond the limitations of the human body. In cyberpunk fiction, the prison of the 'meat' is left behind; in contemporary cyberspace, enthusiasts hope to leave behind sexist and racist prejudices of the meat world. However, as Kirby suggests, the avatars – three-dimensional graphic representations or textual descriptions (depending on the type of MUD) that stand in for the body – that are adopted in contemporary cyberspace unwittingly extend these prejudices by pandering to cultural images of the perfect body.

Cartesian dualism has a misogynistic heritage. The transcendence of pure mind is a position available to the male subject, while the female subject must remain immanent, absorbing all the limits of materiality that man has cast off in his construction of his own subjectivity. An insight regarding the appeal of Cartesian dualism, one that links cyberpunk's rejection of the body to Ross's reading of cyberpunk as male-fantasy wish-fulfilment, is provided by Elizabeth Grosz. She writes: 'It is in this period [puberty] that the subject feels the greatest discord between the body image and the lived body, between its psychical idealized self-image and its bodily changes. Experientially, the philosophical desire to transcend corporeality and its urges may be dated from this period' (*Volatile Bodies* 75). Cyberpunk appeals to the (impossible) desire to escape the vicissitudes of the body and occupy the place of self-mastery. Janet Sayers's work on adolescents suggests that a desire to repress the body is more typical of males than of females: 'men and boys not only describe being disconnected from talking honestly with others in adolescence about their sexual feelings and experiences. They also describe being disconnected from themselves, particularly from their bodies' (88).

The link between adolescent insecurities about bodily control and the cyberspace cowboy's repression of the body is further reinforced by a passage from Gibson's novel. Case, the cyberspace hero, experiences discomfort during an experience of virtually 'inhabiting' another body through the technology of simstim: 'For a few frightened seconds he fought helplessly to control her body. Then he willed himself into passivity, became the passenger behind her eyes ... Case kept trying to jerk her eyes toward landmarks he would have used to find his way. He began to find the passivity of the situation irritating' (*Neuromancer* 56). Case is able to feel in complete control of himself while disembodied in cyberspace, and finds himself irritated by this state of experiencing bodily perceptions without being able to master the material body. Case's interactions with Molly's body provide an interesting twist on the gender politics at work in the novel as well. One of the things that emerges from this setup is that Case has efficacy only in the cyberspace world; it is

Molly who must perform actions in the real world, who has the strength and skills to behave as a typical 'action hero.'

When Molly breaks her leg, she must struggle through the initial pain and then risk permanent damage as she walks out on the broken limb (her pain suppressed with drugs), while Case is able to simply switch away from the pain. As N. Katherine Hayles points out, we are on familiar philosophical ground in this situation, since 'the character immersed in her physicality is a woman and the character who can escape it is a man' ('How Cyberspace Signifies' 118). However, it is also possible to read a reversal of traditional gender stereotypes in this scene: Molly has a physical strength which Case lacks. This reversal is made all the more apparent by the glimpse Case gets of his physical self through Molly's eyes. The narrative abruptly shifts from scenes within cyberspace where Case is 'heroically' using the intrusion program to break T-A security – which seems to mainly involve watching what the program does – to a vision of himself in the physical world as he shifts to Molly's perceptions. He sees 'a white-faced, wasted figure, afloat in a loose fetal crouch, a cyberspace deck between its thighs, a band of silver trodes above closed, shadowed eyes' (256). This vision of Case presents him in typically feminized terms – pale, weak, passive. Further, the language of the passage suggests the degree to which Case is disassociated from his embodied self, since the figure is described as 'it' rather than 'him.' This picture of Case provides an ironic counterview of the cyberspace cowboy as fetal wretch clutching his cyberspace deck security blanket.

Analyses by both Anne Balsamo ('Feminism for the Incurably Informed') and Allucquère Rosanne Stone ('Will the Real Body Please Stand Up?') suggest that gender bias is preserved in 'real' cyberspace as much as it is in cyberpunk representations. Gendered styles of communication and culturally dominant gendered stereotypes of beauty continue to structure people's choices in cyberspace. The repetition of gendered patterns of interaction in cyberspace tends to work to the disadvantage of women, and puts women at risk of material disadvantage as proficiency in technologically mediated interactions becomes more important for employment and social success (Brosnan 33). Further, Stone explains that the technological limitations of rendering bodies in cyberspace using computer technology may actually tend to reinforce stereotypical notions about body, gender, and beauty rather than free the subject from the restrictions of 'meat' judgments. Stone has found similarities between the techniques explored by those working on representing avatars in cyberspace and the discourse of phone sex workers:

'For the work of both is about representing the human body through limited communication channels, and both groups do this by coding cultural expectations as tokens of meaning' ('Will the Real Body Please Stand Up?' 102). Such a representational strategy is more successful deployed in efforts that support the status quo than in those which challenge it.[4] As Rosi Braidotti ironically comments, 'what I notice is the repetition of very old themes and clichés, under the appearance of "new" technological advances.'[5]

Sherry Turkle's work suggests that gendered notions of behaviour as well as gendered constructions of appropriate body appearance are reproduced in cyberspace. Turkle examines the experiences of two individuals who participate in MUDs, a man named, ironically enough, Case and a woman named Zoe.[6] Both tried 'cross-dressing' in their MUD experience, the practice of using an online identity whose gender is opposite to one's own gender. Both Case and Zoe reported that they felt more free to express anger and aggression when in their cross-dressed identity: Case because he felt that an aggressive woman conveyed an impression of strength and sufficiency, while an aggressive man conveyed the image of a bully; and Zoe because she felt that an aggressive man was perceived to be assertive and in control, while an aggressive woman was perceived to be a bitch. What these encounters suggest to me is that cyberspace is not a world in which we can transcend the body and assumptions that are made based on its morphology. This is not to argue that either Case or Zoe did not experience the greater sense of freedom that they reported; instead, it is to suggest that while we are aware of the constructions/constraints imposed upon our own gender, we are not as experienced at decoding those imposed upon the opposite gender. Both Case and Zoe knew how to answer the call of ideology to their material, real-life gender, but both missed the possible readings that may be given to their behaviour based on a presumption of the opposite gender. Additionally, the contrast between these two stories casts significant doubt upon claims that cyberspace cross-dressing provides an opportunity to find out what it is like to 'be' another gender; it merely offers the chance to experience that gender through one's own ideological assumptions about it. Unlike interpellation, this is a model of responding to ideology in which one never forgets that it is a pretence instead of 'really me.'

Cyberpunk thus repeats the typical Cartesian binaries: the male is mind and transcendence; the female is body and immanence. Again, an oft-quoted passage from Gibson confirms this assessment: 'for Case, who'd lived for the bodiless exultation of cyberspace, it was the Fall. In the bars he'd frequented as a cowboy hotshot, the elite stance involved a certain relaxed contempt for the flesh. The body was meat. Case fell into the prison of his own flesh' (6). It should now be clear that elements of misogyny in cyberpunk texts are integrally tied to repression of the body. This being the case, I want to explore the possibility of creating a feminist reading of the social consequences of 'a certain relaxed contempt for the flesh' in our being-in-the-world through restoring the repressed body to my analysis of *Neuromancer.* Despite evidence of Case's contempt for the flesh, both his critics and his imitators have overstated Gibson's rejection of the body.[7] It is thus worth briefly looking at some of the representations of the body in this well-known and much-discussed novel in order to better perceive this gap between the popular image of Gibson's novel and what it actually has to tell us about embodiment.

Despite his contempt for the flesh, all of Case's actions in the novel can be linked to his body and its needs. Case is recruited by Wintermute in the first place because damage to his body has put him in a vulnerable position, cut off from the cyberspace that was both his passion and his source of income. When he was caught cheating his previous employers, they 'damaged his nervous system with a wartime Russian mycotoxin' (6), preventing him from accessing cyberspace again. Wintermute is able to repair the damage so that Case can return to cyberspace, but he adds his own booby-trap of poisons that will reverse the repair unless Case agrees to fulfil his work for Wintermute. While not in cyberspace, Case is addicted to central nervous system stimulants; cyberspace works as a non-biological stimulant that gives him the same feeling of adrenaline and exaltation. Even his desire to transcend his body is rooted in his body's need for a high. Case's situation is an ironic fulfilment of the typical Sprawl[8] joke recounted at the beginning of the novel: 'It's not like I'm using ... It's like my body's developed this massive drug deficiency' (3).

Even Case's eventual emergence from depression and apathy is triggered not by his return to cyberspace, but by the denial of his body's desires: 'the rage had come in the arcade, when Wintermute rescinded the simstim ghost of Linda Lee, yanking away the simple animal promise of food, warmth, a place to sleep ... He'd been numb a long time, years ... But now he'd found this warm thing, this chip of murder. *Meat,* some part of him said. *It's the meat talking. Ignore it*' (152). Finally, Neuromancer attempts to lure Case into a virtual reality world where he can be reunited with Linda Lee – his lover who was killed near the beginning of the novel – in the hope of preventing Case from finishing his mission. Case refuses the virtual world, denying that it is real. Despite his enjoyment of cyberspace, he still insists on a reality based in bodily existence. His love for and connection with Linda cannot be valued if it exists only in the virtual world. Although Neuromancer argues, 'Stay. If your woman is a ghost, she doesn't know it. Neither will you' (244), Case rejects this fantasy and returns to his physical body, following the 'trail' of music playing through headphones.[9]

Case's desire to repress his physical body and live in the world of cyberspace – which is notably a world of disembodied consciousness rather than that of the virtual body offered by Wintermute – is closely linked to a desire to avoid the vicissitudes of the physical body.[10] In a world where everything has become a commodity, even body parts risk being harvested by the more powerful. The cyberspace elite stance begins to look more like a defence mechanism to prevent the subject from realizing how little control it has. Although human action is introduced into the 'other space' of cyberspace, agency within cyberspace is only an illusory escape from the subject's real condition. Alison Adam and Eileen Green suggest that 'the rhetoric of escape – escape from the body, escape from a world gone wrong – has seeded itself into contemporary cyberculture' (95). This juxtaposition of the desire to escape the consequences of having a body with representations of the material consequences faced by non-privileged bodies is one of the ways in which cyberpunk interrogates embodied reality.

Molly, like Case, has had to modify her body in order to obtain employment in this extremely commodified world. Molly has had razor blades implanted beneath her fingernails, and adjustments made to her reflexes, to increase her fighting ability. In order to pay for these augmentations, Molly has worked as a prostitute, a 'meat puppet.' Molly distances her self from the experiences of prostitution: it wasn't 'she' who had sex with the clients, it was simply her body that did: 'Renting the goods is all. You aren't in, when it's all happening. House has software for whatever a customer wants to pay for' (147). Like Case, Molly wants to leave the natural body behind; both try to distance their constructions of self from the actions of their bodies. Both Case and Molly believe that they have agency when they use the body as a technological tool – Case's neural interfaces and Molly's cyborg body – and both feel decentred by the notion of being trapped in the exploitable meat. Although Case and Molly resist the equation of self and body, Gibson counters this perception in his representation of Dixie. Dixie is a ROM personality construct of a former cyberspace cowboy who 'flatlined' during a run. Although Dixie has achieved what is believed to be the ideal of cyberspace cowboys, a completely disembodied existence, he does not welcome this transcendence, but instead asks to be erased once his role in their mission is complete.

*Neuromancer* has served as a paradigmatic text both to SF writers working in the cyberpunk mode, and to the larger community of readers who work within the computer industry. Stone has argued that the novel 'triggered a conceptual revolution among the scattered workers who had been doing virtual reality research for years' and enabled the researchers in virtual reality – or, under the new dispensation, cyberspace – to recognize and organize themselves as a community' ('Will the Real Body Please Stand Up?' 98–9). Gibson's novel articulates a particular type of subjectivity that is interested in repressing the body, and it suggests why this stance would be desirable: the subject wishes to sustain a construction of mastery and the body undermines this construction. Despite the appeal of this fantasy, the body is continually shown to be an inescapable part of Case's subjectivity and the actual condition of being without a body is shown to be an absence of subjectivity. In fact, Gibson himself has said that he dislikes critics who praise his novel for being 'hard and glossy' when 'what I'm talking about is what being hard and glossy does to you.'[11]

The marketing of cyberpunk begun by Bruce Sterling seems to me to have moved away from this insight and toward a glamorization of being hard and glossy, of surpassing the body and its limitations. Contemporary debate on the social consequences of computer telecommunications technology suggests that those readers who identify with Case and perhaps model their subjectivity upon his have created a virtual community that supports Case's vision of the world, if not Gibson's. One of the most commonly expressed concerns is that computer telecommunications via the Internet or virtual reality games have become more 'real' to participants than the real social world, paralleling Case's privileging of cyberspace over bodily existence. Although it is often suggested that telecommunications technologies have the power to create new communities that surpass the limitations of geography,[12] more pessimistic critics fear that the consequence of such activity will be the degradation of the concept of community; we may form connections across the globe while isolating ourselves from our neighbours.[13] As Raymond Barglow argues, new technology-mediated communities will tend to reinforce economic classes and contribute to the gap between those with and without wealth: our sense of responsibility may come to extend to the community of our friends on email, but to exclude the homeless in our own city (202).

Ellen Ullman's *Close to the Machine* provides a personal insight into the subjectivity of someone working in a technically intense job within the computer industry. Her descriptions suggest that the 'elite stance' of the cyber-cowboy has already been adopted by computer programmers, even though we have yet to create neural interfaces that would allow programmers to enter the machine. Although current programmers never leave their bodies even perceptually, the body is still repressed and irrelevant during the act of intense programming: 'Our bodies were abandoned long ago, reduced to hunger and sleeplessness and the ravages of sitting for hours at a keyboard and a mouse. Our physical selves have been battered away' (4). When one works close to the machine, Ullman demonstrates, one begins to think like the machine and adopt the values of hierarchy, consistency, and orderliness that create efficient computer programs. In Ullman's representation, computer programmers view themselves as an elite, more effective than the ordinary users whose poorly articulated and contradictory demands undermine the beauty of the programmers' work: 'In my profession, software engineering, there is something almost shameful in this helpful, social-services system we're building. The whole project smacks of "end users" – those contemptible, oblivious people who just want to use the stuff we write and don't care how we did it' (8–9).

Ullman herself recognizes and diagnoses the danger that such an attitude represents. In allowing everything to become abstracted to data, we lose contact with the consequences that are part of material reality, and we lose the ability to connect our actions to a larger social world. This danger is made evident when Ullman describes a conversation with a fellow programmer who plans to finance his latest project by hosting a porn server: 'the whole complicated business of international pornography had devolved, in Brian's thinking, to the level of a mathematical problem, some famously difficult proof, a challenge of the mind. He seemed neither attracted to nor repulsed by the content of the stuff he would be sending around. To him, it was just bits, stuff on the wire' (62). Ullman describes her own struggle with this realization, as she faces the decision about whether to sell an inherited property on Wall Street. The building is no longer profitable, because telecommunications have changed the commuting patterns of Wall Street workers, and the tenant businesses have lost their customer base. When considering the decision in the abstract, Ullman has no difficulty in choosing to liquidate. However, when she visits the site and is confronted with the material reality of the struggling tenants, she realizes that 'as much as [she] wanted it to become a financial instrument, the building remained solid, material, hopelessly real' (68). Ullman's work is particularly compelling because she acknowledges the attraction of living close to the machine, in a pure and uncomplicated world of data. However, she struggles to retain her own sense of responsibility to the larger social context of her work.

Veronica Hollinger has argued that one of the strengths of Gibson's representation of cyberspace is that it represents social reality as a consensual structure, thereby undermining the power of ideology to naturalize its hegemonic representations. I agree with Hollinger's assertion that 'it is only by recognizing the consensual nature of sociocultural reality which includes within itself our definitions of human nature, that we can begin to perceive the possibility of change' ('Cybernetic Deconstructions' 215). However, the cultural analyses of real engagements with cyberspace suggest that this possibility has not been realized. Instead, as Brian Loader argues, 'for the vast majority of the world's population, the possibility of constructing virtual identities is entirely dependent upon their material situation. Clearly most people are not free to choose but instead are subject to a variety of social and economic conditions which act to structure and articulate their opportunities for action' ('Cyberspace Divide' 10). This fact complements my reading of *Neuromancer*, which suggests that while the novel articulates Case's desire to escape from his material reality, it ultimately demonstrates the futility of such a project. Ultimately, the kind of posthuman that emerges from Gibson's novel is not the infinitely transferable and immortal consciousness of Hans Moravec's *Mind Children*, but instead a being who may have good reasons to wish to escape embodiment but no prospects of actually doing so.

Gibson's vision of a world of cyberspace cowboys who are isolated from meaningful human contact because they treat the world of cyberspace as more real than the material world is one that has been embraced by some as something to be emulated rather than understood as a social critique. Ellen Ullman's reflections on her life close to the machine demonstrate the ethical dangers that this kind of abstraction toward the world implies. Although cyberpunk as a genre has tried to repress the feminine and deny its mothers, feminist critics have insisted upon an excavation of the gender bias that lies at the heart of the genre, and some female writers have brought a female perspective to the sub-genre and its tropes. Most prominent among these is Pat Cadigan, whose cyberpunk novel *Synners* is often seen as the 'feminist' 'answer' to cyberpunk. While *Neuromancer* does show the consequences of forgetting that we are embodied human beings, the novel nonetheless retains a tone of fear toward the body and its vulnerabilities and limitations. Cadigan, on the other hand, argues not only for the necessity but also for the desirability of embodied existence, even in the posthuman age of cyberspace.

Embodied subjectivity is a critical issue for both cyberpunk characters and participants in 'real world' cyber-culture. The critique of contemporary cyber-culture for its tendency to isolate people and destroy the ethical context for actions can be related to feminist criticism of the cyberpunk genre. Jenny Wolmark argues that 'the possibilities for breakdown of identities as part of a transformative social and political process are never realised, at least not in cyberpunk narratives, because the social and temporal experience of cyberspace is centrally concerned with individual transcendence, with escape from social reality rather than engagement with it' (118). Her critique bears an uncanny resemblance to Trevor Haywood's analysis of the social consequences of contemporary cyber-culture. Haywood writes, 'the new monasticism that encourages us to see ourselves as totally independent actors who need no more from life than singular access to the communications port of a computer or an interactive TV is the enemy of truly collective reason and debate, and we must not fall for it' (27). *Synners* is a cyberpunk novel that tries to tell the story another way, emphasizing collectivity and embodied existence.

The novel tells three intertwined stories. The first involves rock music video creators Gina and Mark and their participation in body modification procedures that will allow them to create video directly from the output of their visual cortices. Mark is literally enthralled by this new technology, mirroring Case's attitude of contempt for the meat: 'He lost all awareness of the meat that had been his prison for close to fifty years, and the relief he felt at having laid his burden down was as great as himself. His *self*. And his *self* was getting greater all the time, both ways, greater as in more wonderful and greater as in bigger' (232). Mark articulates the dream of the Cartesian dualist: his self remains intact once freed from the prison of the flesh – in fact, he is about to exceed his earlier capacities. Gina, on the other hand, is more ambivalent about the appeal of socket technology. While she is willing to undergo the procedure, the novel makes it clear that it is her desire to remain connected to Mark, whom she loves, that motivates her decision. Unlike Mark, Gina remains connected to her body, believing that it is a necessary part of being a human subject.[14]

The second story is about Gabe Ludovic, a man who escapes his unfulfilling life of alienating work and a sterile marriage through immersion in virtual reality. In his simulated world, Gabe finds human connection and companionship with two characters that are more real to him than the people in his material world. Cadigan directly links Gabe's desire to escape his body with his desire to avoid the inevitable consequences of decisions that must be faced in the real world: 'too much simulated living, he thought; out here you couldn't just change the program, wipe the old referents, and pick up the story at any point' (388). Cadigan portrays Gabe as someone who has split his own subjectivity, separating himself from those aspects of his subjectivity that are expressed through embodiment, as a consequence of his desire to escape from material reality and his body. Gabe has been cut off from the real world to such an extent that he finds it disorienting to return to embodiment: 'He'd been running around in simulation for so long, he'd forgotten how to run in realife, real-time routine; he'd forgotten that if he made mistakes, there was no safety-net program read to jump in and correct for him' (239).[15] Through the relationship that he develops with Gina, Gabe is gradually returned to material reality and is able to recognize that his addiction to virtual space was related to his isolation from other people: just as Gina believes that engagement with the material world is necessary for ethics, Gabe comes to realize that involvement with other humans has a value that simulated personalities cannot duplicate.

The final story in *Synners* returns to the main motifs in *Neuromancer*: young hacker-heroes and the creation of a self-conscious AI. Cadigan places a female character – Sam, short for Cassandra – at the centre of the hacker community. The community of hackers that Sam belongs to is in hiding from law enforcement because one of them has been caught stealing data about Mark's socket modification. The hackers learn that the sockets may cause strokes, a phenomenon that seems to have developed because Mark experienced a minor stroke while online that is being transmitted to other socket wearers. When Mark has a second major stroke online, one that kills his body but not his now disembodied subjectivity, the stroke becomes a semi-sentient 'virus' – the spike – that infects the entire Net, killing anyone connected by sockets and disabling the communications hardware. Sam and her friends are the only ones who are aware of what has caused the crash of the Net, aided by a self-evolved artificial intelligence named Art E. Fish[16] who exists in the Net. The hackers are able to stage a counter-attack against the spike because they retain an uninfected point of entry to the Net, Sam's modified insulin pump that runs off body power. A connection is established using Sam's insulin pump, and Gina and Gabe use their socket interfaces to virtually enter the Net space and defeat the spike. Like Gibson's descriptions of engagements in cyberspace, they are projected into a three-dimensional environment. Although the form of the body was left unspecified in Gibson's novel, in Cadigan's the connection to material reality is emphasized as the characters inhabit bodily images that match their physical bodies. In this final confrontation, Gabe and Gina are able to defeat the virus by escaping from the simulated reality, a feat they accomplish through a physical connection to one another's bodies in the virtual space, a repeat of the punch that first united them in the material world.

Cadigan's hacker community deconstructs the romanticized cyberpunk ideal that escape from the body is possible or even desirable. Cadigan's characters are more self-conscious about the continuing influence of the body than are Gibson's, in part because the hackers are confronted with Gina and her pragmatic connection to material reality. When a hacker character, Keely, falls down after disconnecting from a near confrontation with the spike online, it is only Gina who looks for an explanation in the material world rather than the cyberspace one: "'Just guessing myself," Gina said tonelessly, "I'd say he fainted from hunger. When's the last time *you* ate?"' (372). Cadigan refuses to let her characters – or her readers – become caught up in the exciting illusion suggested by cyberpunk marketing. Describing the experience of living in an abandoned house while on the run from the police, Sam observes wryly:

> 'But then I started softening up to the idea a little. Thinking that it would be kind of ... oh, exciting, I guess. Romantic, even. Almost like being in the Ozarks again, except freakier. Laptops in the raw, jammers making music. Horny hardware geniuses making cordless modems for you.' She laughed a little and then sighed again. 'But mostly it's being dirty and smelly and not having any safe place to stay and not getting enough to eat.' (268, ellipsis in the original)

In this description, Cadigan demonstrates the appeal of the myth of cyberpunk, but returns her characters – and the community of readers who identify with these characters – to the material facts of social reality.

Cadigan's cyberpunk representations insist upon the material consequences of being excluded from the social community by being labelled
a criminal.

Even Sterling has recently toned down his excessive rhetoric, observing that – to his knowledge – it remains investigative journalists and
police who uncover corruption and conspiracy, not hackers.[17] Cadigan
undermines this earlier heroic image of cyberpunk hackers transcending the body and saving the world. They soon realize that the needs of
the body remain a priority, even when they are working to save the Net
from corruption; as Keely comments, 'Never mind the tech shit, when
do we eat? I wish I'd thought of that when I was busy raiding uninfected
equipment' (376). Cadigan includes within her own cyberpunk mythology Stone's axiom that 'virtual community originates in, and must
return to, the physical. No refigured virtual body, no matter how beautiful, will slow the death of a cyberpunk with AIDS. Even in the age of the
technosocial subject, life is lived through bodies' ('Will the Real Body
Please Stand Up?' 113). The body is so important in this cyberpunk
world that even the artificial intelligence, Art. E. Fish, requires one.
When the Net becomes infected, Art is saved only because his data is
stored in an embodied hard copy, a system of tattoos that record Art's
'self' in a medium safe from the spike.

Anne Balsamo has produced a convincing reading of *Synners*, one
which focuses on linking its representations to real-world engagements
with technology. Balsamo's reading points to the ways in which *Synners*
insists upon embodied rather than escapist solutions to the problems of
engaging with information technology. In both 'Feminism for the Incurably Informed' and *Technologies of the Gendered Body*, Balsamo creates a
template for reading four main characters as examples of four embodied responses to technology: Sam as the labouring body, Gabe as the
repressed body, Gina as the marked body, and Mark as the disappearing
body. Balsamo notes the gendered lines that bifurcate these characters'
engagements with technology: Both Gabe and Mark use the technology
to isolate themselves from the social world and to escape their bodies,
while Sam and Gina use technology to communicate with others, Sam
through the re-establishment of the Net and Gina through her rock vid-
eos. Balsamo discerns a connection between the (typically male) fantasy
of escaping and transcending the body in cyberpunk and the gendered
constructions of the body that circulate in our culture. As Balsamo
argues, Gina's status as marked by race and gender has given her social
experiences that continually return her to a consciousness of her body;

because that body is read by others, Gina, as the subject who is that body, has learned that she must live with the results of such readings. Gina's marked body and her pragmatic insistence that material reality matters return the reader to an understanding of the body as an inevitable component of the subject's social existence. Cadigan, more clearly than Gibson, demonstrates that the desire to transcend the body is an escapist fantasy which is, as her characters ironically suggest, 'only impossible in the real world' (421). As we have seen, Cadigan judges the real world as what matters. The inexorable return of the repressed body is symbolized by the spike – the stroke that Mark avoided by leaving his physical body which now threatens to destroy everything in its path.

Balsamo submits that 'in reading *Synners* as a feminist text, I would argue that it offers an alternative narrative of cyberpunk identity that begins with the assumption that bodies are always gendered and always marked by race.' Cadigan's novel is implicitly informed by Donna Haraway's cyborg politics: the gendered distinctions between characters hold true to the 'cyborgian figuration of gender differences whereby the female body is coded as a body-in-connection and the male body, as a body-in isolation' ("Feminism for the Incurably Informed' 692). I have some concerns about whether the gendered response to technology falls as easily into these lines of connection and isolation as Balsamo suggests. As Balsamo points out, Sam's body provides the necessary ground from which to stage the attack against the spike. What Balsamo fails to emphasize is the fact that Sam's body is able to function as this ground because initially it lacks connection to the rest of the network and so it remains uninfected. Further, the connection to the Net is established also through the socket technology which is associated with two male bodies, Valjean and Gabe.

In the end, it is community and the combination of many people's technological equipment and skills that allow their triumph.[18] Thus, the more important dividing line in the novel is between material connection and virtual connection, not between female-body-in-connection and male-body-in-isolation. The connection to the Net puts them at risk for infection; by herself, Sam could not risk reconnecting to the Net, nor could she effect any change. It is only by combining her 'clean' point of entry with Gabe and Gina's socket implants and Valjean's 'clean' socket interface that the group is able, collectively, to confront and vanquish the spike. At the end of the novel, Gabe has successfully weaned himself of his addiction to virtual fantasy and re-engaged with the material world. However, he isolates himself in a remote ranch house, refusing to have any Net connection in his home. His connections are with geographically proximate people: the local store where he shops, the local schools for whom he provides video production services. Gabe's choice can thus be read as an implicit critique of the risk that cyberspace, in distancing us from present, material communities, will distance us from an ethic of care for others. Gabe refuses to succour his loneliness with virtual engagements, choosing to remain engaged with the material circumstances of his socio-political reality.[19]

Thus, I would argue that *Synners* ultimately transcends the gendered representation of engagement with technology split along body-in-connection and body-in-isolation axes. While Balsamo's reading is certainly a correct assessment of the characters' initial engagements with technology, the novel is finally about the value of human relationships in the physical world, and a caution against allowing all our relationships to be mediated by technology. Cadigan's novel is not anti-technological but, like Donna Haraway's work, it calls for 'a more adequate, self-critical technoscience committed to situated knowledges' (*Modest_Witness@ Second_Millennium* 33).[20] Like Haraway, Cadigan asks us to attend to the material consequences of scientific endeavour, recognizing that 'All *appropriate technology* hurt somebody. A whole lot of somebodies. Nuclear fission, fusion, the fucking Ford assembly line, the fucking airplane. *Fire*, for Christ's sake. Every technology has its original sin. ... Makes us original synners. And we still got to live with what we made' (*Synners* 435).[21] As we learn to live with virtual reality technologies, Cadigan is warning us that it is crucial not to allow them to replace our connection with the material world.[22]

Her cautions can be related to Ullman's description of the ability of data-based abstractions to sever our connection to the social consequences of our actions and choices. As Gina puts it in her description of rock music that attempts to be socially engaged: 'They were all so far away from it, see, they were all so fucking *far away*. They'd say something like "world peace" and they didn't have the first fucking idea of what the world was like. They saved the goddam whales, and they didn't even fucking *live* in the fucking *world*' (198). Ullman puts it more tersely when she observes, 'Surely we were missing something essential if our idea of other people was a program downloaded from the Internet' (181).

The body itself, our most basic connection to the world of materiality and ethics, has also been seen as a technological tool. Silvia Federici argues that the development of a concept of the body as 'human machine' was one of the main social changes concomitant with the rise of capitalism and emergence of labour as a market commodity. She argues that '*the human body and not the steam engine, and not even the clock, was the first machine developed by capitalism*' (146). I want to turn now to a discussion of Mark's refrain 'Change for the Machines' and connect it to this idea of the body or labour as just another machine in the production process.

Initially, Gabe addresses the phrase to Mark, asking him if he needs change in order to purchase something from the vending machines. However, Mark interprets the phrase as a comment on his upcoming socket surgery, and speculates that 'My whole life has been, "Okay, change for the machines." Every time they bring in a new machine, more change' (97). Changing for the machines is a fact of everyday life in our technological culture. Donald Lowe details the changes to physical working environments – and physical injuries on the job – that have emerged from computerization of the workplace.[23] Although changes such as using braces to reduce repetitive motion strain or tinted glasses to reduce computer screen glare are not as invasive as the surgery Mark is to undergo, they do fall into a continuum of modifying the body to more efficiently use machines in our work lives. Performance artist Stelarc believes that 'an ergonomic approach is no longer meaningful. In other words, we can't continue designing technology for the body because that technology begins to usurp and outperform the body. Perhaps it's now time to design the body to match its machines ... What do we do when confronted with the situation where we discover the body is obsolete? We have to start thinking of strategies for redesigning the body' (Atzori and Woolford 197). My reading of Cadigan's novel suggests that she uses the tropes of cyberpunk fiction to argue for the necessity of remaining engaged with a material reality of other human beings. The risk of Mark's desire to change for the machines when considered in the context of this material reality is that we begin to see concrete humans as only another part of a production system, not as something uniquely valuable in themselves. Ernest Yanarella and Herbert Reid in 'From "Trained Gorilla" to "Humanware": Repoliticizing the Body-Machine Complex between Fordism and Post-Fordism' suggest that this is precisely what is happening in their discussion of the 'growing focus on the role of humanware failures (as opposed to software and hardware deficiencies) in these production systems as the main source of injuries on the construction sites' (201).

In changing our bodies to accommodate the use of machines, we change ourselves. In order to use a tool successfully, humans must incorporate that tool into their body image. Even without the physical invasiveness of 'socket' technology, our tools – our machines – become extensions of ourselves: 'The writer would be unable to type, the musician unable to perform, without the word processor or musical instrument becoming part of the body image. It is only insofar as the object ceases to remain an object and becomes a medium, a vehicle for impressions and expression, that it can be used as an instrument or tool' (Grosz, *Volatile Bodies* 80). As Michael Heim has observed, the way in which we incorporate information technology into our body image is fundamentally different from the process for other tools. Unlike other tools, which 'do not adjust to our purposes, except in the most primitive physical sense,' software tools are interactive and change us as we change them, allowing us 'to make any number of tools for different jobs ... The software interface is a two-way street where computers enhance and modify my thinking power' (*The Metaphysics of Virtual Reality* 78).

Computers do not just change our body images, but also influence our thinking and perception; the body and the self are both influenced by information technology. Critics of these changes have pointed to dangers in some of the ways that our subjectivity has been changed for the machines. Ellen Ullman has identified the way that computers can work to change our perception of our environment and other people from a material reality to bytes of information, atrophying our social consciences. Like Heim, she believes that computers are a new sort of tool:

> I'd like to think that computers are neutral, a tool like any other, a hammer that can build a house or smash a skull. But there is something in the system itself, in the formal logic of programs and data, that recreates the world in its own image ... We believe we are making it in our own image ... But the computer is not really like us. It is a projection of a very slim part of ourselves: that portion devoted to logic, order, rule and clarity. (89)

Other critics have also pointed out the damage that changing for the machines can cause. In *The Crisis of Self in the Age of Information*, Raymond Barglow describes the alienating effect that reducing the subject to logic, order, rule, and clarity has on the individual. Barglow is a computer programmer turned psychoanalyst whose work looks at the computer images that are prevalent in the dreams that his patients report to

him. Barglow believes that the computer and the dolphin – both nonhu-
man but sentient beings in his view – are the two current 'mirrors in
which we hope to recognize ourselves' (4). He argues that computers
are like internal mental objects that the child has prior to personal differentiation and identity; they emphasize connection rather than separation and counter the whole notion of sovereignty and self-sufficiency
(6). In exploring the parameters of this self-conception and seeking to
treat his patients, Barglow works in the tradition of ego psychologists
and aims at restoring a sense of unity and self-mastery to a fragmented
subjectivity. Barglow believes that the identifications formed between
computer users and the technology itself contribute to the postmodern
decentring of the subject. Perhaps, however, we need not think of this
decentring of the rational subject as a negative thing. As Balsamo has
observed, the desire to transcend the body is a gendered response of
anxiety that 'signal[s] a desire to return to the "neutrality" of the body,
to be rid of the culturally marked body' ('Forms of Technological
Embodiment' 233); Barglow's crisis of self might similarly be the crisis of
a masculine subjectivity whose centrality and autonomy have recently
been brought into question.

Bukatman has argued, 'the body must become a cyborg to retain its
presence in the world, resituated in technological space and refigured
in technological terms. Whether this represents a continuation, a sacrifice, a transcendence, or a surrender of "the subject" is not certain' (*Terminal Identity* 247). Clearly, Barglow's analysis suggests that changing for
the machines – becoming a cyborg – is a sacrifice or surrender of the
subject who is now in crisis. What remains to consider is whether or not
a feminist reading of this crisis opens up the possibility of a more positive reading of this change. Donna Haraway's influential 'Cyborg Manifesto' suggests that the body/subject who has changed for – merged
with – the machine is an emancipatory figure. The cyborg is a hopeful
figure because, as Balsamo says, 'by disrupting the stable meanings of
the human/machine dualism, other reliable oppositions are also rendered unstable. The cyborg, for Haraway, has the potential to disrupt
the persistent dualisms that have been systemic to the logics and practices of domination of women, people of colour, nature, workers, animals' (Balsamo, *Technologies of the Gendered Body* 35). Unfortunately, this
optimistic reading of cyborg imagery does not seem to be realized in
contemporary engagements with technology. As people change for the
machine, there appears to be an increasing tendency to reduce humans
to the object status of machines rather than to challenge the binaries

that structure our subject/object distinctions.[24] As Ullman observes, the human operator has become just another upgradable component in the computer system: 'the skill-set changes before the person possibly can, so it's always simpler just to change the person. Take out a component, put in a zippier one. The postmodern company as PC – a shell, a plastic cabinet. Let the people come and go; plug them in, then pull them out' (129).

Human workers in information technology are discovering, as did Cadigan's heroes, that 'although *they* have changed for the machines, the machines didn't change for them' (Balsamo, 'Feminism for the Incurably Informed' 692). Both Yanarella and Reid (191) and Lowe (34) observe that the human worker is expected to adjust the pace of production to the standards set by the technology; the machine does not change to accommodate humans. This is true both for the production assembly line and for programmers who work intimately with the machine and, so it would seem, control it. Ullman describes the anxiety of working in a profession which changes as rapidly as does computer technology. As she approaches middle age, she feels anxious in a world that values only what is cutting edge and new; old hardware and old software manuals are merely trash in this world. As Christopher Dewdney has observed, we live in an age of obsolescence: 'we are entering a period of disposable skills, a vast meme landfill of the concepts and routines that we learned for use with various obsolete devices' (25).

All of this change for the machines works to devalue the human and it is this devaluing of human worth that Cadigan seeks to counter with her cyberpunk mythology that privileges real experience over virtual. Barglow argues:

> [T]echnology is essentially contradictory; as a cultural 'text' of a kind, it articulates and extends the fissures and inconsistencies that characterize our lives. We can let that text be written by the interests that currently organize the planet; or we can decide that we are going to write that text collaboratively and democratically, so that technological innovation enlarges the scope of human freedom and self-determination instead of contributing to new forms of irrationality and domination. (182)

Like Cadigan's notion of the 'original sin' of each technology and our need to learn to live with what we have made, Barglow suggests that a political engagement with the deployment of information technology is required to diminish its alienating effects. Cadigan's work does not demonize the technology or suggest that devaluing the human is the only way to imagine either a cyberpunk future or a cyber-culture present. In an interview given at the Virtual Futures conference in 1996, Cadigan said, 'I believe in progress and advances, but I also believe in the soul and I believe that, besides technology being only as good as the people that use it, the culture that springs up around it ... can only be as good to the people that comprise it, as those people are to themselves.'[25] It is not the nature of the technology that determines the effect it will have in its interactions with humans, but the nature of the human culture that creates the context in which a technology is designed and used.

Although both *Synners* and *Neuromancer* explore the appeal of Cartesian mind/body dualism, in the end each affirms that the body is an integral component of subjectivity. Disembodied consciousnesses lack a connection to material reality, and material reality is the space of ethics. In juxtaposing these novels with contemporary critiques of information age culture, I have argued that the novels articulate the same anxieties and dangers that permeate these critiques of contemporary cyberculture. Thus, I see cyberpunk as participating in the critique of contemporary information age culture, providing concrete examples of the risks inherent to life in the information age: cyberspace isolates the individual from connections to other people; repressing or denying the body with technology does not sever the connection between body and subject; and the social world is distorted if our understanding of it is based solely on what is perceived through information technology. The world as it appears in cyberspace is not the complete social world, and cyberpunk cannot offer us a complete cognitive map to this social world; however, cyberpunk can and does offer a map that can assist us in understanding how distorted the perspective from cyberspace is.

Cyberpunk also offers cognitive maps of emerging social formations, and it can intervene positively in the construction of identity performed within these communities. Cadigan's novel thus suggests that the model for becoming posthuman by interface with computer communications technology does not have to be one based on a rejection of the body. Although the desire to transcend both the body and the consequences of a material world is among the founding tropes of both cyberpunk and contemporary cyber-culture manifestations, this need not remain the case. *Synners* shows us that it is possible to imagine a cyberpunk world grounded in the material, in ethical uses of technology, and in respect for other human subjects both on and off the Net. Although the novel dramatizes current cultural anxieties about the need for humans to change for the machines, it also suggests that another possible model is to require machines to change for humans. Computers are mirrors for our subjectivity, not simply in that we look to them for possible models of self, but also in that they reflect the values inherent in the culture that designed them. *Synners* shows us why we must neither forget nor suppress the importance of embodiment as we imagine posthuman selves in an increasingly cyber world.

# 5 Raphael Carter: The Fall into Meat

What monsters show us is the other of the humanist subject. It is the other who must be excluded in order to secure the boundaries of the same, the other who is recognizable by the lack of resemblance.

Margrit Shildrick, 'Posthumanism and the Monstrous Body'

I want to conclude my discussion of embodiment and the cyberpunk genre by looking at Raphael Carter's novel *The Fortunate Fall* (1996), a text which functions as an ironic response to cyberpunk, sharing the sub-genre's tropes without succumbing to its political naïveté. While Cadigan's *Synners* offered alternative subject positions for a cyberpunk age, positions that embrace the material and embodiment both within and outside of cyberspace, Carter's novel goes one step further and insists that we remain focused on the priority of the material as the only reality that 'really' matters. I argue that *The Fortunate Fall* is characterized by its subversion of cyberpunk tropes precisely to restore the repressed body and marginalized material reality to the narrative. This novel functions as an anti-cyberpunk text, explicitly denying the utopian hopes that cyberpunk sensibilities can use information technologies to intervene in the social. *The Fortunate Fall*, then, offers us a vision of what is omitted when we construct a cyberspace posthumanism based on transcendence of the body. The novel's critique of cyberpunk tropes is thus also a warning about the possible consequences of contemporary cyber-culture. If, in Cadigan's terms, we change for the machines by trying to make ourselves into machines, Carter's vision warns us that we will be losing more than simply the limitations of embodiment. Rather, we will be losing ourselves and becoming something less than, not beyond, the human if we accept that self can be separated from embodiment.

The central character of the story, Maya, is a camera, someone who has been modified with prosthetics and nanotechnology so that her experiences can be broadcast to an audience of viewers. The world that Maya inhabits is characterized by extremes of social control. The Postcops – as in Emily Post, guide to socially appropriate behaviour – monitor and 'correct' inappropriate actions in the material world. The Weavers monitor the Net and prevent certain representations from being made. Finally, cameras like Maya work with people called screeners who edit and shape the broadcast, a profession that is somewhere between spin-doctor and censor. Unlike typical cyberpunk narratives, there are no spaces for subversive sub-cultures to hide in Carter's world. Maya's relation to her own body and her desire is controlled through a suppressor chip, which is revealed to be her punishment for the crime of homosexuality. The suppressor chip prevents Maya from feeling sexual desire of any kind, and represses her memory of her lesbian identity.

The story of *The Fortunate Fall* concerns Maya's research into the anniversary of the liberation of the American death camp Calinshchina in Kazakhstan, and her reunion with her lover whom she lost twenty years before when her suppressor chip was installed. Maya's research into the Calinshchina genocide leads her to a character named Voskresenye, a victim of the Mengele-like experiments carried out at the camps. Voskresenye had been a high school student working with a self-organized underground when the American occupation force captured him. He damaged his socket implants by pouring water into them, thereby destroying brain tissue so that the names of his collaborators could not be extracted from his mind against his will. Voskresenye believed that he would die from this damage, but instead the camp's leading 'scientist' devised a cabling link to connect what remains of Voskresenye's brain with the brain of a whale, allowing his damaged mental functions to be restored through their combined mental power. Carter's characterization of the scientist, Derzhavin, shows the limitations of a model of subjectivity that would see humans as extensions of machines rather than vice versa. Derzhavin suffers from the kind of moral attrition that Ullman has argued is a consequence of seeing virtual spaces and data structures as more real than the 'messy' material world. His interactions with others are always mediated by technology such that he starts to see other people as information or objects to be manipulated, rather than as autonomous subjects: 'He seemed to have been born without the gene that enables us to see souls in the world – spirit-blind, as some are color-blind. When Derzhavin looked at a Kazakh or a whale, he saw a wetdisk, an organic computer, sheathed in a husk of irrelevant flesh. The body was an unfortunate complication, and the spirit just a dream of foolish men' (206).

The novel includes a number of characters who have become posthuman by moving their minds from or beyond embodiment in the brain. Voskresenye explains to Maya that the procedure that connected his brain with the whale's has created a single self that is both him and the whale:

> The corpus callosum, the anterior commissure – why, they're no more than a pair of cables; they link the right and left halves of the brain, just as you might link one computer to another. And they're cables wide enough to merge two lobes into one self, so that if you could not dissect, you would not guess the halves were separate. And if you had a cable, well, why not a cable splitter? Could you not set up a cloverleaf among, not two, but *four* lobes? Would they not then be as intimate with each other as the two hemispheres of the brain are? Would they not merge into a single self? (138)

At the climax of the novel, Maya discovers that the person she believed to be her screener – whom she has interacted with only via virtual presence – is actually the lover from her past, whom she had believed dead. The lover – Keishi is the name she adopts in the screener persona, Keiji was her name when they were lovers – explains that she was able to escape death by uploading her mind into the Net, and then returning the 'program' that was her self into a fleshy existence by using part of the whale's mind. Both Voskresenye and Keishi insist that their 'selves' have survived the data transfer intact, but both Maya and the narrative have trouble accepting the notion of the self as a program.

Maya is the epitome – almost the parody – of Foucault's disciplined subject. The cyberpunk utopian notion that we can free the subject by repressing the body and allowing the subject to escape biological constraints is profoundly rejected in Carter's text. Maya's entire identity is erased through the repression of her body and her desire. She not only no longer *acts* lesbian, she no longer *is* lesbian once her mind is cut off from her body's desires by the suppressor chip. When the chip is disabled and Maya's desires and memories are restored near the end of the novel, she resists this change as an erasure of her current self:

But I would not touch those memories. I would keep my muscles clenched around them, I would squeeze them into a ball, hard-shelled and separate. They were in me, but I would not make them part of me; as a sunken anchor does not give up its substance to the ocean, or an acorn passes through the stomach whole. I would not become that other woman, who had died when her lover died, twenty years past. I would remain myself. (263)

In this representation of Maya, Carter overwhelmingly rejects the Cartesian separation of mind and body. Although Maya attempts to refuse to 'become that other woman,' she does not have any choice. When the suppressor chip is disabled, Maya is reconnected to her body as she regains her memory, recalling both her lover's existence and her feelings for this woman.

Although Maya returns to being the woman she was when she was involved in a relationship with Keiji, she refuses to accept that Keishi is truly the person returned. Once again, the body is represented as something that is essential to human subjectivity; Maya refuses to accept that Keiji's personality could have survived the experience of becoming simply data on the Net. Keishi tries to convince Maya that the return to flesh – even the flesh of the whale's brain – has been sufficient to return her soul:

A soul can't live in the Net, no. But there's nothing mystical about it. It's a physical process – for all intents and purposes the soul is serotonin. If you upload your mind to the Net, at least here in the Fusion, you lose your sensory qualia, your emotions; you become a program. But then if you put it back into a brain, the soul grows back. (262)

Maya refuses to accept that Keishi is human, telling her 'You've forgotten what human emotions are like – you either forget them completely, or you blow them up into something they can never be. Damn it, Mirabara, it's only love. It doesn't mean you want to fuse souls with someone. And it doesn't save the world, or even the people in it' (285).[1] In Maya's view, Keishi is only a truncated shadow of a complete person, the type of subjectivity that social critics of the information age fear may become dominant as we move toward understanding humans as analogous to computers. As Ullman has warned, Keishi represents only a very slim part of being human: that portion devoted to logic, order, rule, and clarity.

When Maya and Keishi plan their first meeting in the material world – Maya believing that the possibility to start a new lesbian relationship will exist once her suppressor chip is disabled and Keishi planning to reveal that she is really Keiji – Keishi expresses concern that Maya will be disappointed with the appearance of her 'real' body as compared to the idealized appearance she projects in virtual space. Maya dismisses Keishi's fears, rejecting the idea that the appearance of the physical body is pertinent to love: 'I still don't know exactly how I feel about you, but I doubt it would make any difference if you weighed a hundred kilograms' (175). However, when Maya discovers the truth about Keishi's identity, she rejects any possibility of a relationship because Keishi does not have a physical body. In a macabre enactment of the idea that true love is the marriage of true minds, Keishi wants to move from the mind of the whale – who is dying – to Maya's mind:

> Only a tiny part of me is any kind of flesh. But that part, the part that matters, is in danger. I will die with the whale, unless you let me live. In you … I'll keep my memories on the Net … Everything that's data. I'll just offload a little of your mind into the Net, and take that space. You'll never feel anything missing. But we'll be together. Always. (282–3)

Maya rejects the notion that a relationship conducted only in the virtual space of the Net can be real. In response to Keishi's promises to protect her, love her, and be a companion to her, Maya asks, 'And will you hold me when I'm frightened?' (283). Just as Carter challenges the cyberpunk notion that the subject can be freed by repressing the body, zie[2] rejects the idea that communication technologies can offer new forms of community and new spaces for interpersonal contact. Material reality remains the space of true community and connection.[3]

Another anti-cyberpunk motif in Carter's novel is zir refusal to represent cyberspace as a 'levelling' field in which those excluded from positions of power in the material world can confront the powerful and emerge victorious because of their talents in negotiating cyberspace. In most cyberpunk representations – including Gibson's and Cadigan's – material advantage in the physical world is rendered null when engagements are staged in cyberspace, and the most significant confrontations in the novels occur in this space. Carter offers a corrective to such representations, actively refuting the stereotype of the cyberspace cowboy who succeeds against the odds through personal skill and ingenuity. Maya remembers Keiji's death in terms that specifically reject the cyberpunk fantasy. Keiji had tried to hide their lesbian relationship from public scrutiny on the Net, but eventually they are found out and the Weavers (the Net censor equivalent of Postcops) attack their home. Keiji attempts to attack the Weavers through the Net, forgetting about her vulnerability in the physical world:

> She had spent too much time on the Net, where no situation is ever quite hopeless, and where one person, wired right, can stand firm against a thousand. But I, who had stayed behind to sponge her brow with water, still remembered the inevitability of the flesh ... He looked down along the barrel of his rifle, like one who cocks his head in thought; paused a moment; then lifted his head again and nodded slightly, as though the thought were now complete. A smooth circular hole had been punched in the front window, and another in the wall behind, and between them she lay with a hole the same size in her temple, already dead. (259)

Although Keishi argues that she has survived this exchange – uploading her mind to the Net before the bullet killed her body – Maya does not believe that survival on the Net is true survival. The novel further suggests that their failure to attend sufficiently to material, embodied reality is what has put Maya and Keiji at risk in the first place. During the time that Maya and Keiji were together, Keiji worked to hide their presence as lesbians from representations in the Net, believing this would keep them safe from detection. However, as someone who spends most of her time on the Net, Keiji has seemingly forgotten that they live in the material world, and they are discovered. When Keiji is trying to figure out what mistake she made, she tells Maya, 'They must have followed me in reality, that's all I can figure. I'd hidden you from the Net so well I thought they'd never find you – after all, who thinks of *Weavers* wearing out actual shoe leather, and all that Sam Spade kind of shit?' (271). As opposed to most cyberpunk fiction, the 'real' action in this novel happens in the material world, not cyberspace. Keiji and Maya are discovered – and destroyed – by material-world forces that are more powerful than they and that do not care about Keiji's virtual prowess.

What I find most interesting about *The Fortunate Fall* is the connections it draws between representation, subject formation, and ideological hegemony. Technological surveillance and control have created a world of utter repression and utter stability. There is no space for the articulation of reverse discourse in this world: the Weavers monitor and control discursive representations on the Net, and the Postcops appraise and restrain performative representations in the material world. The body modification of sockets – which allow Maya to work as a camera, and allow her viewers to access her feelings and experiences over the Net – means that the mind as well as the body can be disciplined into models that are deemed appropriate. Suppressor chips ensure that even one's thoughts and bodily desires can be examined and moulded by hegemonic ideology.

The novel explicitly argues that controlling representations allows one to control the bodies and subjects who will materialize in the social world.[4] Voskresenye believes that it is necessary to make public representations of things in order for the things to be real. Something that cannot exist in discourse cannot exist at all. He believes that the censorship exercised by the Weavers is 'being used to enforce an official vision of humanity' (236). Voskresenye broadcasts Maya's experience of de-suppression as the memories and desires repressed by the chip return, arguing that the discursive representation of Maya's lesbianism will make space for lesbian identities to be lived in the material world: 'Because of what you did today, there may yet come a time when they no longer have to hide' (261). The story of Maya and Keiji suggests that discursive suppression is effective in controlling subjectivities. Maya describes their relationship as 'roach love, furtive and opportunistic, scattering at the touch of light' (255) and Keishi justifies the invasion of Maya's privacy via the broadcast by arguing, 'there had to be a world for us to live in. You know what happened last time – how it wore us down, how we could only live by hiding' (277). This notion that repressing representations of subjectivity can also curtail its material expression has its antecedents in the history of repression of gays and lesbians in our material world. In her review of the laws against homosexuality, Jane Ussher recounts the resistance to naming lesbianism in discursive practice, even in the legal discourse for prosecutions of lesbian behaviour. Following Jeffrey Weeks, she links this resistance to a fear that naming lesbian activity in discourse would produce lesbianism in women who would otherwise never have considered an alternative to heterosexuality.[5]

Voskresenye suggests that discursive representations are overwhelmingly important in a social context in which information technology mediates social interactions. Maya asks him why the horrors of the Nazi holocaust are remembered while the far larger casualties of Calinshchina are forgotten. He believes that a change in the technological context of representation is the key to understanding the difference:

'We are like men forced to walk about in darkness,' he continued, 'except in one chamber where our eyes are uncovered. If the color blue were not found in that chamber, we would never know that it existed; and if in the chamber all men were well-fed, we might forget that there is hunger in the world. The chamber would impress itself upon us so forcefully that nothing else seemed real. And so it is. Telepresence is a chamber in which a new sense, more important than sight, is uncovered. What happens outside the chamber barely exists. And so you see, if what we call reality is to persist, *everything* must be brought into that chamber.' (231)

Voskresenye's speech links Carter's novel to critiques of the power of information technology to reduce the material world to an abstraction of data. The more that communications technology mediates our social interactions, the more our constructions of social reality will be filtered through the representations provided via the technology.

Critics of information culture have suggested – as does Voskresenye – that the social community of the Internet is not representative of the entire range of human behaviours and identities. Voskresenye argues that 'the Net *should* be the most democratic form of communication that the world has ever known. It *should* replace the poor bumbling of human compassion with perfect electronic sympathy – instant, universal understanding, available to everyone' (236). In the novel, the reality turns out to be that the Net is used to impose a narrow interpretation of human normalcy on a diverse population. In fact, Keishi suggests that the reason that the human soul cannot be uploaded into the Net in the Fusion[6] is because censorship of the Net excludes the full range of information that is humanity: in Africa, a Net without Weavers, souls can be uploaded (68). Allowable representations control how (much of) the subject materializes. While the language of the soul has unfortunate spiritual connotations that link it to Christianity as an organized religion, the concept of the soul in the novel seems decidedly different, primarily because the soul is so clearly an embodied, material entity in this world, not an abstract spiritual essence. Looking beyond the disquiet raised by the connotations of the language, we can see Carter's notion of the soul as an important aspect of any vision of the posthuman. Like Octavia Butler's warnings that a radically reduced range of human variety at the level of genetics may produce an inadequate kind of posthumanism, Carter's idea of the soul as that which cannot be uploaded into a censored system of representations cautions us about what will be left out of a cyber-culture posthumanism. In both cases, the problem is not with the technology in and of itself, but with a use of this technology that posits current ideological assumptions as the only possible reality.

Virtual realities far more overtly control the materialization of bodies that matter and those that cannot materialize. As a tool of representation, virtual reality threatens to create a posthumanism that works with a definition of baseline normal humanity that is as restrictive and oppressive as the possibilities imagined by genetic engineering. Voskresenye believes that he is restoring sin to the Net by violating Maya's privacy and broadcasting her de-suppression experience, but he argues that sin – the full range of human behaviours and desires – is a necessary part of social reality. Voskresenye believes that the distance between preventing exploitative representations and repressing those whose values differ from our own is too short:

> But observe how easy a descent it is, Maya Tatyanichna. First, viruses that control minds; certainly we don't want those. Then, feelings so intense they might cause damage to the audience. Then, things which simply disturb people. Finally, anything which might be a bad influence – for after all, if you control the world-soul, anything that you exclude does not exist. (234)

He realizes that his actions will destroy the stability of the current social configuration, and that the return to chaos will be a negative thing in many ways. However, he argues that social stability is not worth the cost of social conformity. Diversity must be allowed, even though it generates conflict: 'I would not permit a Utopia built on the backs of the one per-cent, of the few remaining dissidents, even those who no longer know what they are' (244). The title of the novel refers to the Christian myth of expulsion from the Garden of Eden, the fall into knowledge of good and evil. Carter presents this fall as fortunate because it is also the fall into free will: if both good and evil exist, humans have the freedom to choose an ethics. The fortunate fall is also the fall into the body, the fall from the impossible cyberspace ideal of disembodied existence into the material reality of the flesh. Like the expulsion from the Garden of Eden, the fall into the flesh comes with a price as well as a reward. As we saw most clearly in *Neuromancer*, the fall into the flesh is the fall into vulnerability. However, the fall into the body also produces the space of ethics.

Material reality provides an ethical ground for our decisions while the abstractions of cyberspace allow us to detach our actions from their social context and thereby obscure their connection to ethical consequences. Voskresenye admits that in his action he has been guilty of the same evil as motivated the Guardians in their experiments at Calinsh-china, the willingness to sacrifice concrete individuals to abstract ideas: 'that is what it *means* to be a Guardian: to think that individual rights are a dangerous folly, and compassion merely sentiment. The greater good is everything – and a greater good not to be measured empirically, but defined ideologically' (244). However, unlike the Guardians – and some users of cyberspace as described by Ullman – Voskresenye never forgets or discounts the concrete individual, Maya, whom he has sacrificed to an abstract ideal, freedom of representation. And ultimately, this is what ethics is about, not making perfect or painless choices, but accepting the consequences, both good and bad, for those we make.

The necessity for ethics of both embodiment and attention to material reality is a vision that emerges very clearly in *The Fortunate Fall*, but it is also implicit in both *Neuromancer* and *Synners*. Although the cyberpunk sub-genre has a reputation of repressing and denying the body, a close reading of these three texts does not support this position. My choice of texts in constructing this argument is not accidental. Each marks an important moment in the discursive construction of the sub-genre: Gibson's text as the 'first' cyberpunk narrative; Cadigan's as the 'feminist' response to cyberpunk; and Carter's as a self-conscious refutation of cyberpunk-as-subversion arguments. My reading has foregrounded the ways in which these cyberpunk novels offer a critique that coincides with that made by social critics of information technology culture. Stone contends that the context of cyberpunk fiction has contributed to creating a sense of community for those who work in virtual reality technologies; that is, that representations in the novels are self-consciously internalized and enacted by such subjects. Her observation suggests that one of the risks connected to attempts to intervene in the social construction of reality through the discourse of fiction is that aspects of the work – like the abstractions of cyberspace – can be taken out of context. I believe that this is what has happened in the reading of Gibson's novel by those researchers who – as Stone argues – have formed a sense of community based on the novel. While contempt for the meat can be found in cyberpunk discourse, this perspective is limited to certain characters in the novels, and a reading of it within the context of each novel suggests that this perspective is not endorsed.

Returning the repressed body to the received understanding of these texts can intervene in the identifications that reading subjects make with characters in the novels. The novels themselves comment on one another and suggest that SF writers, too, are offering enhancements of and corrections to earlier representations and readings. Cadigan's novel represents the importance of the material body much more forcefully than does Gibson's, and her novel can in some ways be considered a refutation of Case's fantasy of bodily transcendence. Carter, in turn, forces readers to acknowledge that disciplinary forces of social control are at work in the cyber-world as much as they are in the material world, refuting fantasies that suggest that one can escape one's social position by creating a new one in cyberspace. By drawing attention to the importance of the body in each of these texts and demonstrating how it remains an integral component of subjectivity in them, I hope to draw attention to the continued importance of embodied existence in our contemporary world.

The line between science fictional representations and non-fictional social projections is blurring. In his analysis of contemporary culture, *Last Flesh*, Christopher Dewdney argues that we are living in a time of 'transition between the human and posthuman eras' (2); his book is entitled *Last Flesh* because, he argues, we may be the last generation to be limited to an embodied existence. Like Lee Silver's work *Remaking Eden*, which I discussed in chapter 2, Dewdney's blurs the lines between fiction and social commentary.[7] *Last Flesh* is filled with the imagined lives of posthuman citizens as they check on their children via telepresence robots (109–11), change their biological sex, or enhance their genitals (164–6). Dewdney discusses the work of Hans Moravec, the director of the Mobile Robot Facility at Carnegie Mellon, who argues in *Mind Children* that there are no insurmountable scientific obstacles to uploading human consciousness into computer systems. Once again blurring the lines between science fiction and science popularizing, Dewdney follows this discussion with an imaginative description of experiencing the uploading process (172–4). Following this description, Dewdney suggests that 'at this time we cannot know if we are dependent upon embodiment or not. Disembodied consciousness might be insufficient to maintain sanity. Or it might be a liberation of sorts ... and we will fly further and faster on our wings of thought than we had ever dreamed possible. We may well find out' (175).

While Dewdney's rhetoric offers the possibility that disembodied consciousness may not be a 'sane' way of being in the world, his representational trope of flying faster and further than we thought possible invokes cyberpunk representations of the freedom and exaltation of life in the matrix. Identifying the subject with a disembodied consciousness and treating the world of cyberspace as more meaningful than the material world entails reductions that obfuscate moral facets of our social being. Returning the repressed body to a critical engagement with cyberpunk fiction suggests that we are dependent upon embodiment for our moral being. The unexpected other that emerges from my reading of cyberpunk fiction is the body itself; presumed to be irrelevant, it nonetheless returns to remind us that we are embodied beings, engaged in a material context. It is imperative that we do not lose sight of this fact in our engagements with information technologies. The Möbius strip image of the body as part of the subject – not disposable flesh – is where we must begin, or we risk cyberspace's becoming a social space that marginalizes those who have been 'reduced' to the body by discourses in the past.

My conviction that the appeal of cyber-culture, like cyberspace, is rooted in a desire to escape responsibility for ethical actions in material reality was reinforced by reading the November 1999 issue of *Shift*, a magazine targeting those 'living in digital cultures.' I purchased this magazine to get some sense of the degree to which cyberpunk representations are being reproduced in discourses aimed at self-identified cyber-culture citizens. My reading of the magazine confirms many of the fears that cyberspace works to distance people from real community and tends to produce an attitude in which others are treated as objects rather than subjects. The magazine includes an article on cyberdildonics or cybersex,[8] describes the various feedback prostheses that can be purchased to create a virtual sexual experience,[9] and provides information about hiring an online partner if required.[10] The emphasis on pornography on the Internet and the materialization of a body via sexual prosthetics suggest that the body is being returned to cyberspace, not in representations that demonstrate the embodied nature of subjectivity, but only as a 'meat puppet.' Cyberpunk texts that emphasize the materiality of the body and the relationship between embodied reality and ethical action are a necessary supplement to such depictions. An advertisement for the video game *Duke Nukem*, also appearing in this issue of *Shift*, suggests that the desire to escape the ethical aspects of embodiment may be part of the appeal of cyberspace. 'In real life,' reads the advertisement, '*Duke Nukem* would be forced to attend "sensitivity training." Real life sucks' (*Shift* 25). I have no doubt that *Duke Nukem* and other video games advertised in 1999 have become obsolete before anyone has a chance to read this paragraph. However, the fact that this short-lifespan computer technology includes video games only emphasizes the concerns about obsolescence of 'components' that move only at a 'human speed' that I discussed in the previous chapter. This rapid turnover in relevant references is yet another example of how we are changing for the machines.

Andrew Ross has argued that, in their engagement with technology, cultural critics need to develop 'something like a hacker's knowledge, capable of penetrating existing systems of rationality that might otherwise be seen as infallible; a hacker's knowledge, capable of reskilling, and therefore of rewriting the cultural programs and reprogramming the social values that make room for new technologies' ('Hacking Away' 100). This hacker-life knowledge must return the repressed body to our discursive engagements with technology and the subjectivities that are forming around its use. It is imperative to rewrite the cultural codes and reprogram the social values that have excluded the body from these discourses, and return to them a sense of ethical grounding in material reality. Finally, I believe that it is important to continue to offer critical assessment of cyberpunk representation – such as the one I have provided here – to counteract those readings of the texts which tend to glorify the suppression of the body represented by characters such as Case. As Shannon McRae points out in 'Coming Apart at the Seams: Sex, Text and the Virtual Body,' the difference between representing something and endorsing that which is represented is sometimes lost in reception. She writes, 'Gibson's paranoid vision of a world rendered nearly uninhabitable by multinational corporations, whose hegemony is enabled by means of a vast, interlinked information network, started out, like most good science fiction, as social criticism. Now it has become the model upon which various corporations, keeping up with enormous consumer demand, are carefully planning and busily constructing a brave new world' (242). Cultural critics must strive both to offer other visions of our new information technology world, and to counter such misreadings of already existing ones.

Just as cyberpunk is a misleading map of the social world, the cyberspace cowboy is a misleading model for the posthuman. Gibson's novel provides us with a vision of the limitations of this figure, ending with Case left alone once again, lacking a connection to anyone or anything once the contract with Wintermute is over. Although he has longed for human connection through his fantasies of Linda Lee, the truth is that he never had the ability to form a meaningful bond with anything other than his cyber deck.[11] Sam, in *Synners*, moves us closer to a viable model for posthumanism. Although she remains engaged with the cyber-world

and technology, she also remains rooted in material reality and connected to other humans. Unlike Case, Sam is part of both a community
and a family. Carter's novel, finally, suggests that the problems with this
model of posthumanism lie not only with the cyberspace cowboy, but
with the very design of the matrix itself. Through Voskresenye's insistence upon returning sin to the Web, Carter reminds us that the body
stands in for many aspects of humanity that are repudiated by models of
posthumanism, such as Hans Moravec's, that suggest that the self can
become a code. The visions of the cyber posthuman, like those of the
genetic posthuman, must recognize that what needs to be transcended
in a move from humanism to posthumanism is not the human body but
instead the narrow vision of humanity that has been characteristic of
humanism as a discourse.

# 6 Jack Womack and Neal Stephenson: The World and the Text and the World in the Text

The body is indeed the privileged object of power's operations: power produces the body as a determinate type, with particular features, skills, and attributes. Power is the internal condition for the constitution and activity attributed to a body-subject. It is power which produces a 'soul' or interiority as a result of a certain type of etching of the subject's body.

Elizabeth Grosz, *Volatile Bodies*

I have been examining the way in which discourses about 'appropriate' and 'inappropriate' uses of technology both authorize and prohibit the materialization of particular body-subjects through their competing representations of the body. In this chapter, I want to turn to the technology of writing itself. Marshall McLuhan has called language 'the first technology by which man was able to let go of his environment in order to grasp it in a new way' (57). I have been examining the ways that the discourse of popular fiction intervenes in the social construction of subjects and arguing that it can provide a space for the social formation of subjects that runs counter to dominant ideology. This chapter provides a reading of two science fiction texts which themselves examine the processes by which reading and writing texts shape the body-subject. Both Neal Stephenson's *The Diamond Age* and Jack Womack's *Random Acts of Senseless Violence* foreground the practices of writing and subject formation: *Random Acts* is in the form of a young girl's diary, and narrates her practice of writing a new self that can cope with her changed social circumstances; *The Diamond Age* describes the interactions of a young girl with the automated *A Young Lady's Illustrated Primer*, an interactive book that responds to the girl's social circumstances and offers stories to teach her how to negotiate her social space.[1]

These two texts emphasize the relationship between disciplining of bodies, internalizing of cultural images, subject formation, and cultural resistance. My argument in this chapter is that the representations of subject formation as mediated through texts provided in these two novels model how cultural texts – including popular fiction – can be related to subject formation. In particular, in reading these two novels I want to demonstrate how they problematize the relationship between the ideological call and each individual's response to that call. Both novels demonstrate that the social context of the addressee of the call produces variations in his or her response to it. Such variations open the space for agency and resistance that Judith Butler has theorized in her idea of variation on repetition, so that citations both re-produce and resist/alter the dominant ideology. The effect of a cultural representation cannot be known in advance or 'programmed' in a deterministic way; the material matters – which bodies matter matters – because it is one's position within the material structures of power that will shape one's access to cultural products and one's response to their ideological calls.

Both novels focus on the perspective of subjects at the margins of power, offering responses to the bourgeois world of the reader in *Random Acts* and the neo-Victorian culture of the *Primer*'s creator in *The Diamond Age*. The world of Stephenson's novel is divided into a number of enclaves or tribes which are centrally governed by the Common Economic Protocol. Nell, our protagonist, is a thete, someone who lives in the Leased Territories rather than belonging to a particular economic enclave. Nell's life is characterized by under-privilege: she spends her days entertaining herself and fending off the sexual and physical abuse offered by her mother's stream of boyfriends. Nell never leaves her apartment, and her brother Harv provides what care and education she is given. The New Atlantans, a neo-Victorian culture, are one of the most powerful tribes. Nell's life intersects with the New Atlantans when Harv steals a copy of the interactive book *A Young Lady's Illustrated Primer* from its creator, John Hackworth. Hackworth is unable to report this theft because, although he designed the software, he is also in possession of the book illegally.

Hackworth created the *Primer* as a commission from Lord Finkle-McGraw, an equity lord who believes the New Atlantan culture is becoming stagnant because its citizens lack the creative spark of subversion. The *Primer* is to be his gift to his granddaughter, one he hopes will provide this missing element to her socialization. Hackworth steals a copy of the *Primer* for his daughter, Fiona, with a similar wish that it will allow Fiona to rise above her given station in life and achieve some financial equity (the enclave's marker of upper from lower class). Hackworth believes the *Primer* is required for Fiona to achieve this goal because the socialization of New Atlantans – aimed at producing stability – inhibits them from taking risks. He observes that 'he'd met a few big lords ... and seen that they weren't really smarter than he. The difference lay in personality, not in native intelligence. It was too late for Hackworth to change his personality, but it wasn't too late for Fiona' (81).

Although Fiona does eventually get her copy of the *Primer* through other means, Harv steals the first illicit copy produced by Hackworth and gives it to Nell. Access to the *Primer* literally changes Nell's life, demonstrating one of the main themes of the novel: that access to knowledge is one of the most important formative elements in one's life. A judge from the Celestial kingdom, another powerful tribe in whose courts Harv's case is heard, reflects on how to punish the theft of the book and observes, 'a book is different – it is not just a material possession but the pathway to an enlightened mind, and thence to a well-ordered society' (163). Through the *Primer*, *The Diamond Age* shows that popular culture is not a deterministic monolith. The message embedded in a cultural text, its hegemony, is articulated differently in interaction with different readers. The *Primer* is

> a catalogue of the collective unconscious. In the old days, writers of children's books had to map these universals onto concrete symbols familiar to their audience – like Beatrix Potter mapping the Trickster on to Peter Rabbit. This is a reasonably effective way to do it, especially if the society is homogeneous and static, so that all children share similar experiences. What my team and I [Hackworth] have done here is to abstract that process and develop systems for mapping the universals onto the unique psychological terrain of one child – even as that terrain changes over time. (107)

The *Primer* uses live ractors (interactive, virtual reality actors) to produce the speech coming from the book. The particular way that the *Primer* functions as an interactive technology provides an insight into the way that the same cultural representations are received, taken up, inhabited, and used differently in different social contexts. There are three copies of the original *Primer* in the novel belonging to Nell, to Hackworth's daughter, Fiona, and to Finkle-McGraw's granddaughter, Elizabeth.[2] Although each girl starts out with an identical database of cultural information, the stories that the *Primer* tells them are different because their social circumstances are different.

The *Primer* has a much more dramatic effect on Nell's life than it does on either Elizabeth's or Fiona's. The reason that the *Primer* takes on such importance for Nell is that the knowledge it embodies and passes on to her is knowledge to which she would not otherwise have had access given her class. The *Primer*

> sees and hears everything in its vicinity. As soon as a little girl picks it up and opens the front cover for the first time, it will imprint that child's face and voice into its memory – ... And thenceforth it will see all events and persons in relation to that girl, using her as a datum from which to chart a psychological terrain, as it were. Maintenance of that terrain is one of the book's primary processes. Whenever the child uses the book, then, it will perform a sort of dynamic mapping from the database onto her particular terrain. (106)

The story that the *Primer* tells Nell recounts the adventures of Princess Nell and her escape from imprisonment by her evil stepmother in the Dark Castle. Nell's harsh social circumstances give her story the darkness of Grimms' fairy tales rather than the lighter tone of much children's literature. For Nell, the *Primer* becomes a tool of survival, parenting her in the absence of her neglectful mother.[3] The *Primer* reassures Nell when she becomes lost, it teaches her to prepare nutritious food for herself, and it teaches her self-defence. The *Primer* is an interactive text, so reading it falls somewhere between reading fiction and participating in virtual reality. Although Nell never is immersed in another environment through technology, the *Primer* does tailor its stories to her questions and to what it can 'see' of her surroundings. Nell's willingness as a reader to translate between text and life is crucial to the success of the *Primer*. For example, Nell learns her self-defence skills through practising the motions she sees modelled in the *Primer*.

The *Primer* works with the intention to shape Nell as a social subject. At the beginning, the *Primer* tells stories to Nell based on its observations of her social circumstances. The *Primer* initially offers solutions to the problems Nell is experiencing. However, as Nell gets older, the *Primer* increasingly requires her to drive the narrative by finding solutions to problems on her own. The *Primer* works to shape Nell's behaviours into models that are appropriate to the set of values that form its programming. For instance, it begins the process of teaching Nell to read by insisting that she tell her dolls bedtime stories. Noticing her activity of tucking the dolls into bed, the *Primer* intervenes and announces, 'For some time Nell had been putting them to bed without reading to them … but now the children were not so tiny anymore, and Nell decided that in order to bring them up properly, they must have bedtime stories' (95). The *Primer* gives Nell knowledge that she would not otherwise be able to gain from her limited social experience, defining any terms in its stories that are unfamiliar to Nell and branching off into sub-stories to explain concepts when necessary. In addition to the skills that the *Primer* teaches Nell, it also teaches her self-reliance and perseverance through its pedagogical style. For example, during a story about camping in the woods, Princess Nell is provided with flint and shown that the flint can make sparks, but Nell the reader must work through the logic of starting the fire with dry leaves, blowing on the flames, etc. The *Primer* allows Nell to repeat stories as many times as she wants so she may learn from her mistakes and continue trying until she succeeds.

The primary role of the *Primer* is thus not to instil the knowledge of cultural archetypes into the young girl who reads it, but to shape her subjectivity along the lines of the values embodied in the neo-Victorian response to these archetypes which is embedded in its programming. In Nell's case, the goal of the *Primer* from the first day that she reads the story is to remove her from her current social surroundings. The story of the *Primer* remains the same from the first day Nell reads it, although nuances and details continue to multiply as she gets older and more sophisticated. The story that the *Primer* tells is about how Princess Nell is able to escape from the Dark Castle, but how she must leave Harv behind because he is 'too big and had to stay locked up' (109). The *Primer* thus prepares Nell for the day when she will leave the Leased Territories, and for the fact that Harv will not be able to follow her into her new setting. The *Primer* produces the result it predicts when Harv and Nell are forced to run away from home to escape their mother's extremely abusive current boyfriend. The children seek refuge with Brad, one of their mother's former boyfriends. However, only Nell is able to stay with him because her speech and comportment patterns – learned from the *Primer* – allow her to blend into the enclave to which Brad belongs. As the *Primer* predicted, Harv is 'too big' to escape with Nell, in the sense that his social being, as the author of criminal acts, is already written, and this social being precludes him from entry into the enclave.

Given the radically different social context between Hackworth (the programmer) and Nell, the social perceptions that the *Primer* gives Nell do not always prove reliable. Although the *Primer* teaches Nell not to trust strangers through a story in which 'the ractive [was] made in such a way that, once she'd made the decision to go away with the stranger, nothing she could do would prevent her from becoming a slave to the pirates' (225), it is Harv who must teach her 'You can't see 'em. They don't look like pirates, with the big hats and swords and all. They just look like normal people. But they're pirates on the inside, and they like to grab kids and tie 'em up' (67). Similarly, when Nell and Harv run away from home, Nell is drawn to a park as a place of refuge, since Princess Nell spends a lot of time in enchanted forests. The children must rely on Harv's real-world experience of their socio-economic context to understand that they will be attacked by security drones if they remain in the park, since those who own it refuse to let transients take up residence there.

Eventually, the *Primer* successfully interpellates Nell as a subject who can negotiate the world of the New Atlantans. Sponsored by Lord Finkle-McGraw – who has discovered the theft of his commission and who desires to see how his theories of its efficacy will play out in Nell's life – Nell attends a school in New Atlantis. However, it is when Nell is immersed in the New Atlantan social context that the differences between her early life experience and those of the girls born into this enclave become apparent. This difference between the representations of the *Primer* and Nell's social experience is the source of her ability, ultimately, to resist the subject position the *Primer* prepares her for, and to strike out to 'seek her fortune' (469). The gap between experience and representation, I will argue, is the space for the agency of the social subject to resist the call of ideology and to articulate a reverse discourse.[4]

In contrast to Nell's social trajectory from poverty to privilege, the character of Lola in *Random Acts of Senseless Violence* follows the path from privilege to social exclusion. Although it is set in the near future, *Random Acts* lacks the technological elements that are expected in a SF novel.[5] Other novels within Womack's series make it clear that it is not that technological innovation is absent from this world: it is just that the experiences of Lola, the twelve-year-old protagonist, do not include access to this technology. The world in this novel shares some of the characteristics of the cyberpunk world such as a belief that governments will increasingly come to lack power as corporations gain it, and that individuals will become increasingly irrelevant to the economy, which will dominate social life. Where Womack's text differs radically from cyberpunk representations is in the status of the body. Lola is not a hacker striving to escape the meat of her body. Instead, Lola's body is central to her subjectivity and to the narrative as she strives to ensure its physical safety in her increasingly violent social circumstances, and as she struggles with her 'queer'[6] sexual desire.

This novel clearly reveals how the irrelevance of the body is a subject position available only to the privileged. The novel narrates the changes in Lola's life and character as she moves from occupying the subject position of daughter of middle-class parents to occupying that of home-less street-gang member. The picture readers receive of this world is limited to the perspective that Lola is able to provide, which entails an absence of technology and a failure to imagine the source of political power. In contrast to the hacker-heroes of many cyberpunk narratives – often teens who live away from their families in communities of hackers – Lola is not concerned with infiltrating data stores or acquiring the latest deck. Instead, as her daily life becomes increasingly full of violence and risk, Lola's focus is on her physical survival. Her survival is not predicated on her ability to master technology but on her ability to use her body to protect herself (and eventually to act out her aggressions against the world).

The novel is written in the form of Lola's diary. The book opens in February when Lola receives the diary from her parents as a present for her twelfth birthday, and ends in July when she abandons the diary – and any attempts to continue her previous social existence – to join the DCons, reputedly the most dangerous street gang. Lola names her diary Anne, recalling to the reader the diary of Anne Frank and the similarly abrupt and violent ending of her narrative.[7] One of the main themes of the novel – and another connection to the story of Anne Frank – is the violent, material consequences for social subjects who are othered by oppressive political power structures, as happens to Lola and her family. *Random Acts* demonstrates in a visceral way the difference between occupying a body that matters and occupying one that does not. Theodore Schatzki argues that a change in social circumstances – a change to the disciplinary power that acts upon the body – will produce a new social subject: 'Merely subjecting a body to particular conditions suffices to produce persons of new types' (54). Lola's story epitomizes this argument. Lola's body and how it is positioned in the discourse of others are crucial to her identity throughout the novel and to her growing sense of isolation, which eventually erupts into violence. When her family is forced to move to the 'ghetto,'[8] Lola is rejected by her private-school friends for inhabiting a poor body; she is later further rejected by both her friends and her sister for inhabiting a queer body; finally, she is rejected by the new friends she makes in the new neighbourhood because she inhabits a white body.

Lola's growing sense of isolation is paralleled by events in the external social world of the novel, a world that the reader is able to see only through Lola's eyes. What glimpses we do get of the larger social context suggest a world in recession that is increasingly polarized into us versus them categories. The news reports are filled with stories of riots in various cities, while the teenage prank of setting homeless people on fire no longer even makes the news. The government turns the army on its own inner-city citizens in Operation Domestic Storm, while Lola's fundamentalist Aunt Chrissie and her husband are 'buying semi-automatics because they think there'll be an uprising of the maids and gardeners' (35).

Lola and her family's increasing poverty and their need to occupy the social space of poverty fragment their perception of social reality and open a gap in the hegemonic discourse, a discourse they previously saw as natural. Lola reports the following exchange with her mother while watching the President on the news:

> He said the nation was poised for recovery like he always said. He also said on the advice of advisors mobs of animals in the cities would be shown no mercy. 'Does he mean us?' I asked Mama. 'No sweetie he means everybody else' Mama said. We watched some more but he didn't say anything else but how great America was. 'How will they tell the difference between us and everybody else?' I asked but Mama didn't say anything like the answer should be obvious. (120)

Lola more readily adjusts to the fact that they have moved to the new neighbourhood, and that their social position is not the same as it used to be, than does the rest of her family. Her father is increasingly absent at his new job, her mother is increasingly absent in a world of anti-depressants and anti-anxiety medications, and her sister retreats into sleep and obsessive rocking. Only Lola acknowledges their new social circumstances and engages with the people in the new neighbourhood. Her experiences of living in the neighbourhood allow her to gain the critical perspective that – other than to the still-privileged outsiders – the differences between 'us' and the President's 'mobs of animals in the

cities' is not an obvious one. Unlike her mother, Lola no longer believes
that there is a distance between her family and everyone else in the
neighbourhood. The rhetoric of Othering deployed from the perspective of privilege – the perspective represented by the television – can no
longer perform its ideological work on Lola; its obviousnesses are no
longer obvious.

Lola's mother tells her that they gave her the diary to record everything that happens to her so that she 'could remember how sweet life is
even when it doesn't seem like it anymore' (9). The diary becomes for
Lola a kind of autobiography in which she charts her changed social circumstances and writes herself a new self that is capable of surviving in
this new context. In his study of the process of constructing narratives in
autobiography, Mark Freeman has observed that Saint Augustine used
his autobiography to write for himself the better self he wanted to be,
and then, finally, to translate this discursively articulated self into his
material self.[9] Lola's diary functions in a similar way, but instead of writing the self she wishes to become, she writes of the increasingly violent
self she finds herself becoming in response to her changed social circumstances. Lola articulates her desire to avoid becoming this discursive
self, and sets up an opposition between writing and acting such that her
ability to express her frustration to Anne works to prevent this frustration from erupting as violence in the material world.

Lola's diary charts her attempts to remain 'the same person' she was
when she lived in her old neighbourhood and her inevitable failure in
this project because the social institutions and people around her insist
upon reading her in a new way once her material circumstances have
changed. Lola indicates in her writing to Anne that she initially feels out
of place in the new neighbourhood, but that necessity aids in her adjust-
ment:

> Anne where Jude lives is so awful but after a while I got used to [the neigh-
> bourhood] and it felt like a nice place. It's like our apartment now it's not
> as good as our old one but it's home just the same. It was weird though that
> you could adjust to something so quick and I wondered if I were Jude if I
> could ever get used to living in a place like hers. (125)

Lola adjusts because she must. Although her parents insist that the
move is only temporary, she is aware that they are lying; she comes to
understand that they are lying to themselves as much as to her: "'we
won't be gone that long I'm sure of it" Mama said. "I believe you" I said.

"I'm sure of it darling oh don't worry" Mama said. Then I knew she was trying to convince herself and not me so I didn't say anything else' (58). The changes to the material and disciplinary setting of Lola's body produce changes in her subjectivity. Lola understands that she is changing and feels her 'real' self slipping away. She writes, 'I'm worsening when it comes to writing Anne but my energy drains too quick sometimes ... I paid the phone bill but rent's coming due. I can't remember what I used to be like Anne it fears me' (231). Lola has moved from the material existence of twelve-year-old private-school girl to that of a street-gang member who must now worry about how to steal the rent money; it is not surprising that Lola has difficulty remembering her previous identity or that her changed material conditions have produced a new interiority. Writing the diary is Lola's attempt to articulate a self that she can still recognize, but as the material conditions become more and more demanding of her time and energy, she is less able to write to Anne.

One of the things that Lola articulates clearly in her diary is her increasing sense of isolation. Anne becomes the one friend who will listen to her and understand her need for community. Lola becomes a criminal, a social outsider, in large part in response to the ways in which she feels society has excluded her and her family, producing them as bodies that do not matter. Lola personifies Anne as a correspondent who will notice gaps in the conversation and who may consequently become annoyed or worried. She apologizes and explains when other demands prevent her from writing on a daily basis: 'Anne I'm sorry I'm not being as diligent as I should be in writing to you every day but you know how it is with distractions like parents and sisters and school' (26). Lola has learned from her parents the strategy of denying or minimizing problems, and she therefore feels that Anne is the only safe outlet for expressing her own anxieties. When she comes into conflict with her private-school friends over her queerness, she writes to Anne, 'I didn't want to talk to Boob because she wouldn't understand and I didn't want to talk to Mama because I didn't know what she'd say. I wanted to talk to you Anne but Wednesday night it was hopeless' (98).[10] Not only does Lola feel that she can't talk to her family about her own problems, but she also feels that she must try to help them solve their problems. Anne becomes the only 'person' who is available to Lola rather than making demands of her: 'Sometimes I think no one in my family needs a shrink they just need me. I need somebody but outside of you I don't know who Anne' (137). Eventually, however, simply writing to Anne is not sufficient to provide Lola with a sense of community. As she tells Anne, 'nobody else hears when I word except you problem is you don't talk back' (219).

Lola increasingly articulates the need for her diary to vent her frustrations as an antidote to acting them out in the material world. Lola's diary becomes a catalogue of all the ways in which she and her family are disadvantaged by their economic fall, and of her sense of helplessness to solve any of the problems. Her sister becomes progressively catatonic as the private-school girls attack her for her newly perceived difference from them. Lola reports:

> After we moved here they started calling her Ghetto Girl even though we're not ghetto people we just live in a poor neighborhood. They wrote welfare mother on her desk in Geography class in black magic marker and her teacher made her wash it off. They laugh at her clothes even though they're the same clothes she was wearing before we moved here. (165)

Lola's father is exploited at his new job as manager for a bookstore, forced to work overtime without compensation and to pay for any mistakes employees make under his supervision. This exploitation culminates in the owner's refusal to give the family the last paycheque, because the overtime required to cover for Lola's father after his death 'evened out' (216) with the amount of money owed to him. Lola vents her sense of helplessness to prevent or fix her family's suffering as rage in her diary. She writes 'So I'm just miserable Anne and I'm so mad but there's too many people to be mad at. Sometimes I think I'm going to go post office [crazy] like everyone else' (111).

Lola contains her desire to 'go post office' by using the diary and her sense of Anne as sympathetic addressee to control her feelings. However, as Lola comes to feel increasingly socially isolated, the diary proves less and less efficacious for containing her rage. After she is forced to fight with another girl to prove her membership in the neighbourhood gang, she tells Anne, 'when I think about what she did I get so mad I can't even think. That's not good because one thing I've always prided myself on is that it used to be I can think when everyone else is going crazy. It's getting harder and harder though. I just feel even more alone than before' (144).

Increasingly, Lola does not have the distance to reflect on her options in the diary before choosing a course of action. The immediacy of violence and danger in her new surroundings forces her to react rather than think. Recounting a confrontation between gang members she says, 'I couldn't believe how violent they got so quick it was scary' (142), and, reflecting upon her own attack on a man who grabs her friend Iz on the street, she worries about her lack of concern in attacking another person:

> Today I suddenly felt bad about hitting that guy like I did when he jumped out and put a hold on Iz. Anne what's the matter with me why didn't I think of it before now? Sure he could have hurt her and when I think about it I know I'd do it again if I had to but why did it take so long to rack me like this? It wasn't like he wasn't human or anything. I could have killed him and it took me till today to care. (180)

Initially, the violence in Lola's diary is metaphorical and used to express her feelings in a controlled manner. She explicitly links the process of writing to her containment of acting: 'I have to write and tell you what happened today because I feel like I'm ready to explode if I don't' (91). However, as she becomes increasingly isolated from her original social circumstances and increasingly absorbed in the day-to-day life of her ghetto neighbourhood, the diary is no longer sufficient to contain her anger. The street language that she begins to use instead of her original prose suggests this need for action. The street language displays a pattern of verbing[11] most nouns, producing a sparse prose style that emphasizes immediate action.[12]

The breaking point for Lola comes when she feels that her new friends in the street gang have rejected her. Lola has consoled herself with the thought that, although she is cut off from her previous social existence, her connection to these new friends ensures a continuing social space for her. However, two incidents challenge this belief and sever Lola's last connection to a larger social network. In the first, Lola calls Iz and asks her to come to the hospital when Lola's mother collapses. Another friend, Jude, who has been a competitor with Lola for Iz's affections, arrives instead. Jude tells Lola that she is making too many demands on Iz and that Iz needs to put herself and her own people ahead of Lola. Lola believes that Jude and Iz are rejecting her because she is 'girl exclusive,' while both Jude and Iz sleep with boys as well as girls.[13] Jude informs Lola that it is her race rather than her sexual preference that forms the basis for their rejection of her: '"Girl that's not what's meant" she said. "We're tribal. You're not. Her flinging with you's a catkiller nada more. Am I incoming or not?"' (238).

After this incident, Lola recounts her sense of isolation to Anne:

I'm unsoulled Anne I'm racked total now constant. I never felt so lone
bereft lifelong. It's an evil year when everybody skips me first my Brearley
friends then Daddy Boob and now Mama and Iz gone gone the same day.
Rethinking what's already downgone my aching breaks me open like I'm
bleeding everywhere an allover visit from granny. We lived right one time
Anne and then it all popped there's no knowing why there's not. What did
I do to bring down this what. When I solo now I feel constant set to blow
like I could bloody everbody I see unreasoned I know but that's that. I
don't see how it's handleable but everybody bypasses somehow they say but
it's hard to think I will there's too much I hate now. What did I do Anne
what did I do. (240)

Cut off from everyone, Lola feels anger that is not 'handleable,' anger
that can no longer be contained in her diary. The final break for Lola
comes when she is separated from Iz and Jude in a riot. Jude calls for
rescue from some influential sexual clients, and a limousine comes to
save her and Iz. Lola watches the limousine run over people in its path
as it leaves the scene and later accuses Iz, 'You'd driven over me too if
I'd been in front of you' (250). Lola's escalating desperation is reflected
in her prose style, which moves from relatively correct grammar in the
earlier diary entries to a spill of run-on sentences, comma splices, and
fused sentences as the book progresses.

After this break with Iz, Lola moves to break with the one 'person' left
in her life, Anne. Her narrative suggests that, in her desire to avoid
being rejected by this final 'person,' Lola chooses to reject Anne before
Anne can reject her. Lola writes herself out of her book, having com-
pleted the discursive articulation of her new, street self and therefore
moving on to its material existence outside of the text. Now alone and
'unsoulled,' Lola decides to take material action against someone who
has wronged her and helped to produce her as this abject subject. She
describes her rage to Anne, not as a strategy for containing it, but in
order to share her plan for acting it out with her friend. Lola again
experiences the sense that she can no longer recognize the self she has
become, telling Anne, 'When I eye myself mirrored I don't see me anymore it's like I got replaced and didn't know it but I'm still here under-
neath I'm still here' (241). She translates her violence from discourse
into action by attacking Mister Mossbacher, the owner of the bookstore
where her father worked, and beating him to death with a baseball bat.

Lola is unable to perceive the larger power structures at work that

have produced both her and Mister Mossbacher in their respective subject positions, but Womack suggests a social world in which those who
hold real power are beyond the reach of abjected individuals like Lola.[14]
Mister Mossbacher is within Lola's sphere of influence, but killing him
does not do anything to change the institutional structures that disem-
powered Lola and her family. After attacking Mister Mossbacher, Lola
gets a glimpse of this larger, determining structure when she notes, 'It
weirded me sudden that Mister Mossbacher owned a store but housed
in a building like ours' (252), but she is too far gone in her rage to care.
Mister Mossbacher does not even know who she is or why she kills him.

After crossing this line from discourse into action, Lola realizes that
she can no longer return to being the old self who began to write the
diary. Her final entry is her good-bye to her friend, Anne, as she abandons her efforts to construct a self through narration and goes to join
the DCons gang. The DCons function metaphorically in the text, marking a line between those living by the old rules and those who entirely
reject even the new social order and are in revolt against it. By commit-
ting murder, Lola indicates her readiness to become one of the DCons,
to leave her old self and its community connections behind forever.
Lola's final diary entry – a mere six months after the diary began – concludes: 'Can't cut me now. Can't fuck me now. Can't hurt me now. No
more. No more. Night night Anne. Night night. I'm with the DCons
now' (256). Writing on the role of victimized characters in postmodern
narrative, Mark Ledbetter argues:

> Victims in an ethic of reading and writing are those persons desperate to
> be heard and seen (note passive tense) and whose alternative to a literal
> disappearance from the human story is to commit desperate acts of vio-
> lence to themselves, even to those they love, in order to create a world that,
> while not of their choosing, is at least of their making. (22–3)

Lola chooses to destroy the remnants of her old social world – living with
her family, trying to make friends in the neighbourhood – by crossing
the line and joining the DCons rather than waiting for these things to be
stripped from her by outside forces. The violence that Lola offers to the
world through her attack on Mister Mossbacher is a reflection of the violence she believes the world has offered to her as it gradually stripped
her of all human connection. Her defiant final diary entry, 'Can't fuck
me now. Can't hurt me now,' suggests that, ultimately, she is more victimized than victimizing. In the final analysis, Womack's title is ironic: the

acts of violence are neither random nor senseless. Instead, acts of violence are deployed within a social context in which one cannot reach the true author of one's abjection and so one attacks those within reach, targets who perceive the act as random and senseless. Lola's diary is her resistance to this characterization of her violence, insisting upon the rationality of her target and the necessity of her act. She offers the diary as a counter to the distortions of official narratives, claiming, 'I started penning cause you had to know what happened today it's what they'll never tell and memories don't flypaper everything' (249).

Both *Random Acts* and *The Diamond Age* suggest that culture, not nature, forms the social subject, and both posit the technology of reading and writing as central to this social shaping. *The Diamond Age* opens with an epigraph from Confucius which reads, 'By nature, men are nearly alike; by practice, they get to be wide apart.' This insight structures the novel and Stephenson's representation of the role of the *Primer* in shaping social subjects. The *Primer* is created because of Finkle-McGraw and Hackworth's belief that culture creates the difference between people who will become exceptional and those who will remain marginal: 'Finkle-McGraw began to develop an opinion that was to shape his political views in later years, namely, that while people were not *genetically* different, they were *culturally* as different as they could possibly be, and that some cultures were simply better than others. This was not a subjective value judgment, merely an observation that some cultures thrived and expanded while others failed' (20–1). Miss Matheson, the head of the neo-Victorian girls' school that Nell attends, explains to Nell that superior culture is the foundation of the New Atlantan society:

> Now, there was a time when we believed that what a human mind could accomplish was determined by genetic factors. Piffle, of course, but it looked convincing for many years, because distinctions between tribes were so evident. Now we understand that it's all cultural. That, after all, is what a culture is – a group of people who share in common certain acquired traits. (321)

Finkle-McGraw and Hackworth believe that the ideology embedded in the *Primer* will produce its reader as a social subject who embodies all that is strong about neo-Victorian values while at the same time contributing an element of independence – of the subversive – that will produce the reader as a creative and innovative thinker who can expand on the culture of her tribe. The title of the *Primer* brings to mind Victorian conduct books, interpellation machines that are designed to produce the same result in each reader. The interactive quality of the *Primer* – its ability to perceive and respond to the environment of its reader – results in the *Primer* being a different text that forms a different subject for each reader. Finkle-McGraw offers his view on this relationship between text and context when he observes Nell, Fiona, and Elizabeth playing together. He says to Hackworth, whose daughter, Fiona, sees the book as a fantasy escape world:

> I will expose myself to the risk of humiliation by predicting that Elizabeth reaches the wall first; that Nell finds the secret way through; but that your daughter is the first one to venture through it … Elizabeth is a Duke's granddaughter, accustomed to having her way, and has no natural reticence; she surges to the fore and claims the goal as her birthright … Nell stands above the fray and thinks … To the other girls, the wall is a decorative feature, no? A pretty thing to run to and explore. But not to Nell. Nell knows what a wall is. It is a knowledge that went into her early, knowledge she doesn't have to think about. Nell is more interested in gates than in walls. (289)

While each of the girls has been taught to explore her environment by the *Primer,* their various social experiences beyond the text have produced them as radically different social subjects in their method of engaging with the world.

Elizabeth learns from the *Primer* to question authority and the given, but her protected social circumstances do not provide her with a goal toward which to direct her energy. Elizabeth eventually runs away from the New Atlantan tribe and joins CryptNet, a tribe that embodies anarchy and secrecy, but which has no social goal on which to focus its subversive energies. Fiona treats the *Primer* as a fantasy space, a realm in which her father – who gave her the *Primer* just before leaving for a ten-year secret mission – is not absent. To Fiona, the *Primer* is an alternative to her material existence and an escape from its vicissitudes. Fiona eventually leaves the New Atlantan tribe and joins the Dramatis Personae, a troupe of actors whose style purposefully blurs the lines between acting and living. It is only Nell, for whom the *Primer* has provided a means to change her given social circumstances and create a space for herself in a different social world, who treats the *Primer* seriously, as a tool of social engineering.[15]

Nell eventually learns from the *Primer* how to use discourse as a tool to shape worlds of her own design, both in her work of writing fantasy scenarios for Madame Ping's brothel, and in her eventual leadership role for the new tribe of citizens produced by the second edition of the *Primer*. This second edition of the *Primer* is produced en masse – 250,000 copies – to serve as the primary education tool for a group of refugee Asian girls who have been saved from death after being abandoned by their parents because of famine and their gender. The second edition of the *Primer* uses a voice generator rather than live ractors to speak to the girls. It is static, not addressing them individually but telling them Nell's story as it has evolved to the point at which it was copied. By the end of the novel, the other tribes are forced to acknowledge that these girls 'constitute a new ethnic group of sorts, and that Nell was their undisputed leader' (489). The extremely isolated social experience these girls had while growing up means that the *Primer* has been the only influence on their social constitution. The Asian girls therefore do not make any distinction between their loyalty to Princess Nell in the story and their loyalty to Nell as leader in the material world. This unique cultural context has produced a new social subject position.

*Random Acts* also suggests that it is culture, not nature, which produces social being. This theme is present in *Random Acts* primarily through the narration of Lola's transformation into a killer through her experience of social exclusion. Additionally, *Random Acts* suggests that the difference between the private-school girls and the street-gang girls is access to education, not superior intelligence. When she visits Jude's home in an abandoned building, Lola notices a certificate on the wall that says 'Outstanding Student Sixth Grade Judy Glastonbury' (124). Lola also learns that Jude no longer goes to school because she had to run away from home to escape her rapist father. At the beginning of her diary, Lola announces 'Boob and I love school' (12), but after her experiences of ostracism by her classmates, Lola tells Anne, 'I hate it that it's school again tomorrow and that's sad because I used to love school so much' (148). The influence of culture in producing the social subject is largely seen to be the effect of the way in which one is positioned in the representations of others, and treated materially as the subject one appears to be in such discourses. Both the teachers and the students start to treat Lola and her sister differently after they move. It is this sense of how others view her that contributes to Lola's increased hatred of school and therefore to her social production as a delinquent. Lola tells Anne, 'I wondered why I was even bothering [to go to class], since no one would talk to me once I got there and I was being treated like I was dysfunctional or a challenged child or something' (110).

Lola is struggling with the changes brought about by her family's economic downfall at the same time as she is dealing with her own developing sexuality. Lola continually fears that the reason that she is different from everyone else, the reason that she goes 'post office,' is because she is queer. Womack, however, makes clear not that Lola's sexual desire sets her apart from everyone else, but that her social abjection – in part because of this desire – produces her as a criminal subject. Lola's struggle to come to terms with her queerness parallels her attempts to think through the social abjection of ghetto subjects. Initially, although Lola feels sexual desire for her friend Katherine and they kiss, both insist that they 'just kissed because we wanted to' (87) and not because they were queer. Lola understands that the position of queer is a socially abject position and so she insists that her sexual desire for girls must be something else, because she does not perceive it to be pernicious in the way she understands queer is supposed to be. Later, Lola comes to acknowledge her desire to herself, but insists upon keeping it hidden for fear of social rejection.

Although she believes Jude and Iz have 'been queer together sometimes,' she tells Anne, 'I didn't say anything else because I didn't want Iz to think I was a queer if she's not' (134). Finally, after a sexual experience with Iz, Lola begins to reject the negative construction of queerness because it doesn't fit her experience:

> Anne it was so nice just lying in bed with Iz holding and petting each other like kittens. If I closed my eyes it was like I was touching myself but I wasn't I was touching Iz which is a weird feeling but good. Maybe I am queer Anne but if I am it's not awful that's all. (185)

Lola continues to worry that the difference of her queerness is the source of her social isolation, writing, 'There's nothing wrong with me there isn't but everyone else thinks so when they know even if they don't know' (204). Lola recognizes that other people think there is something wrong with her desire – 'everyone else thinks so when they know' – but argues that this judgment is based on a misperception of what that desire actually is – 'even if they don't know.' Finally, Lola insists to Jude, 'Nada's wrong with me cause I girl exclusive' (238).

Jude's response to Lola – that it is because of her race rather than her sexual orientation that she is positioned as an outsider – ties in to the parallel between discursive constructions of queerness and discursive constructions of ghetto subjects. Lola responds to Jude, 'Maybe it's true that what's blooded tops all but if so it's a worse world than I ever specked Anne that limits who's close overmuch and divides and conquers just like the big boys want. Love's love whoever's loving Anne it always seemed to me' (238). This response functions equally as a condemnation of racism and of homophobia. It is Lola's love for Iz that helps her to understand that the pejorative construction of ghetto subjects is as false to the real material existence of people living in the ghetto as the pejorative construction of queer subjects is false to Lola's experience of sexual desire for Iz.

Lola reflects on the reasons why her friends at her private school will no longer talk to her and decides that 'It couldn't be just because we had to move up here but maybe so, I've heard them talk about public school girls like Iz before and it's always like they were just deadhead trash. Maybe that's what they think I am now' (113). While telling Anne about a conversation between her right-wing aunt and her mother, Lola begins to deconstruct the discursive boundary between her neighbours as ghetto people and her family as people who just live in a poor neighbourhood: 'Chrissie told her that if we're murdered in our beds Mama has no one to blame but herself since she insists on keeping us here surrounded by what Chrissie calls those people. Iz is those people and I love Iz she's my best friend' (182). Once she experiences poverty, Lola understands the falsity of representations of poor people and their reasons for living as they do. The absurdity of Lola's conflict with one of her teachers over her appearance suggests the breadth of the gap between experience and representation: 'Today Miss Wisegarver asked me why I'd stopped dressing up for class and wearing jeans and khakis instead. I told her we'd had to cut back on our dry cleaning which is the truth. She shook her head and said that anyone who really cared about how they looked could always find the money for dry cleaning if they really wanted to' (112). Lola's next comment in her diary recalls the scarcity of food in the house, indicating that she is able to reject the teacher's statement, seeing it as a biased perspective rather than as the 'truth.'

These examples of homophobia, classism, and racism suggest that one of the ways in which the social produces its subjects is through abjection, and that this process of exclusion is responsible for the very result it fears. Both *The Diamond Age* and *Random Acts* portray a social world in which group loyalties are the most important determinant of fate. This emphasis on the importance of community can be linked to the critique of liberal humanism as the model for posthuman identity. Versions of the posthuman which put the unique individual and his (possibly her) freedoms at the centre of their moral program entail a corollary emphasis on personal responsibility. While the individual is free to pursue whatever might suit his or her fancy – with the proviso that he or she not curtail the freedom of others – he or she is also free to fail, and this failure to thrive will be deemed a consequence of the subject's own inadequacy. Further, in its emphasis on personal freedom over community, this liberal humanist version of the subject reinforces a social order which obscures the distribution of power by institutional structures and which allows successful individuals within it to believe that their success is a 'natural' expression of their personal superiority.

In opposition to this perspective I argue that the success of the social subject is not an expression of personal merit or individual will, but is instead a measure of the degree to which that subject occupies a socially validated subject position. The community will support or abject the subject based on where that subject falls on the map of the culturally intelligible, and how the subject is interpellated by this social community will produce a corresponding interiority. *Random Acts* also clearly rejects a Cartesian view of the subject as associated with mind and reason alone. As the passages I have quoted reveal, the novel is intensely emotional and it not only shows Lola's emotional response to her situation but also appeals to the reader on an emotional as well as intellectual level as we read Lola's anguish. It is precisely her isolation and lack of community that lie at the heart of Lola's outpourings of anguish to Anne, reinforcing the limitations of the isolated and autonomous individual as a model of subjectivity.

The key difference between Nell and Lola is that Nell is able to move into a community that will protect and value her, while Lola loses the various communities in her life (private school, family, neighbourhood friends) as the story progresses. Nell's teacher Miss Matheson tells her:

> It's a wonderful thing to be clever, and you should never think otherwise, and you should never stop being that way. But what you learn, as you get older, is that there are a few billion other people in the world all trying to be clever at the same time, and whatever you do with your life will certainly be lost – swallowed up in the ocean – unless you are doing it along with like-minded people who will remember your contributions and carry them forward. That is why the world is divided into tribes. (*The Diamond Age* 321)

It is Lola's lack of any tribe, any community, that produces her as a

'mindlost' (*Random Acts* 252), abjected subject able to offer only violence. Lola continually insists that she fails to recognize herself in this
angry person, arguing that she has been produced this way from the
outside in rather than from the inside out. After killing Mister
Mossbacher, she tells Anne, 'what got me most was that my hands
looked like somebody else's it's true when that's said' (252). Lola's diary
can be read as a kind of autobiography. As Mark Freeman has theorized
about narrative voice in autobiography, 'the narrator, rather than being
the sovereign origin of what gets said, is instead a kind of passage
through which those discourses presently in circulation speak' (198).
What is being spoken through Lola's voice is the way in which the gap
between (individual) experience and (social) representation can produce an abject social subject. Abject subjects are often silenced subjects,
ones whose voices cannot be heard in ideological debate because they
do not speak from a position of legitimated power. Lola's story warns of
the danger of such silencing, and how such subjects may become forced
to turn to actions when their words are ignored.

The rapidity of the change that Lola experiences both personally and
socio-economically makes her example an extreme one. The *Primer* provides a more appropriate model for how a great deal of social shaping
functions on an ongoing basis rather than during crisis. Nikolas Rose
has theorized that democratic societies rely on discourses that encourage social subjects to shape themselves in socially advantageous ways as
a method of ensuring social stability without overt repression. Rose suggests that such societies encourage people to think in certain ways
through 'professionals and experts not only flourishing within the appa-
ratus of state but also promulgating their visions of how to identify and
solve problems through the sale of their expertise on the market, and
through the dissemination of their messages through the industry of
mass communication and popular entertainment' (122). Rose argues
that liberal democracies depend on such indirect mechanisms 'through
which the conducts, desires, and decisions of independent organizations and citizens may be aligned with the aspirations and objectives of
government not through the imposition of politically determined standards, but through free choice and rational persuasion' (122).

Cultural ideological apparatuses, such as popular fiction, work to
encourage their readers to identify with particular types of characters
and particular types of choices and hence produce social subjects that
correspond to the values promoted by that fiction. Rose's work focuses
on the power of the 'psy' disciplines – psychiatry, psychology, and their

various popularizations – to encourage the subject to perform the work of monitoring and disciplining the self. The *Primer* functions as Rose argues the 'psy' disciplines do. It teaches Nell appropriate courses of action by narrating the consequences of poor choices. Further, it encourages Nell to monitor her own behaviour, to 'stand above the fray and think,' as Finkle-McGraw puts it, through its interactive operation, which requires Nell to make choices in order to move the story forward.

As a social interpellation machine, the *Primer* should be flawless. Although the novel shows us only Nell's *Primer* in action, we can assume that the other first edition copies function in the same way, since they are constructed from the same code. The *Primer* not only includes a database of appropriate cultural knowledge, but it tailors its use of that knowledge to the situation at hand. Further, it writes the girl who reads it as the protagonist of the story, encouraging a blurring of boundaries between discursive self and material self. Finally, the *Primer* functions as an interactive medium, teaching the reader skills she can apply to the material situation at hand, and requiring her mediation to continue the story. Given all the strengths the *Primer* demonstrates as a tool of social subject formation, why is it not successful in producing identical subjects from each of the three girls who read it? The answer, I suggest, lies in the fact that the social context for each girl's experience of reading is radically different. It is this gap between discursive representation and personal experience that suggests both the limitations of texts in their function of disseminating ideology, and the possibility of a space for agency in their reception.

Nell learns the lesson about the difference between representation and experience early in her life through the *Primer*'s story of the evil Baron Burt. Baron Burt – a character based on Nell's mother's boyfriend Burt – emerges in the story in response to Burt's physical abuse of Nell and Harv. In the story, the character of Dinosaur, based on one of Nell's dolls, plans to kill Baron Burt with a long, sharpened pole after Princess Nell and Harv have tricked Burt into drinking until he passes out. In the material world, Burt also drinks until he passes out. Rather than run away, as the *Primer* suggests, Nell decides to imitate the character of Dinosaur. In her previous readings of the *Primer*, she had been able to learn by imitating characters other than Princess Nell in the story, gaining the skills possessed by these characters through this imitation. Nell creates a long screwdriver in the matter compiler[16] and attacks Burt while he sleeps. Her attack only awakens and enrages Burt and the children must run away from home in order to escape him.

Nell reflects upon this incident when she is older: "'I have been angry at my *Primer* ever since that night," Nell said. ... "I cannot help but feel that it misled me. It made me suppose that killing Burt would be a simple matter, and that it would improve my life; when I tried to put these ideas into practice..."' (281, second ellipsis in original). The Constable, a guardian of the Victorian enclave, responds:

> Now, as to the fact that killing people is a more complicated business in practice than in theory, I will certainly concede your point. But I think it is not likely to be the only instance in which real life turns out to be more complicated than what you have seen in the book. This is the Lesson of the Screwdriver, and you would do well to remember it. All it amounts to is that you must be ready to learn from sources other than your magic book. (281–2)

Nell takes the Lesson of the Screwdriver to heart, and begins to apply the critical reading skills she has learned from her interaction with the *Primer* to her reading of the social world. The crucial lesson that Nell learns from this conversation is to question representations rather than to take them as a straightforward reflection of reality. The Constable teaches her to understand this difference as that between education and intelligence. Nell comes to realize that while the *Primer* can educate her, she needs to reflect on her experiences both within the *Primer* and within the material world in order to derive any intelligence from them. The Constable tells her that 'the difference between stupid and intelligent people – and this is true whether or not they are well-educated – is that intelligent people can handle subtlety. They are not baffled by ambiguous or even contradictory situations – in fact, they expect them and are apt to become suspicious when things seem overly straightforward' (283).

After this exchange, Nell begins to find more subtlety and ambiguity in the experiences of Princess Nell in the *Primer.* As Nell nears the end of the story, all of the tales in the *Primer* have to do with Turing machines. Each of the various castles that Princess Nell encounters in her journeys has a social organization that is controlled by a Turing machine. The *Primer* teaches Nell how Turing machines work and how to change their programming by requiring her to solve various problems at each of the castles, problems related to Turing machine malfunction. The *Primer* has learned from Nell's conversation with the Constable, and Nell discovers that 'in recent years the *Primer* had become much subtler than it used to be, full of hidden traps, and she could no longer make comfortable and easy assumptions' (346). Through her experiences with Turing-machine-constructed social worlds within the *Primer*, Nell learns to view the various tribes that make up the material world as additional examples of Turing machines. As with the castles in the *Primer*, Nell is able to understand social structures as the outcome of the systematic application of rules, that is, as arbitrary albeit orderly constructions. Her position as an outsider who has experienced life in contexts other than the neo-Victorian one allows her to understand neo-Victorian values as one programming choice among many rather than as 'natural' or 'right.'

More importantly, Nell's ability to perceive the material world as another Turing machine provides her with the insight that the social order is not only constructed, but also subject to change. Once she understands the neo-Victorian society as an institutional structure programmed to work within a defined set of rules, she realizes that one can control the outcome of the 'program' by understanding its parameters: 'She had the neo-Victorians all figured out now. The society had miraculously transmutated into an orderly system, like the simple computers they programmed in the school. Now that Nell knew all of the rules, she could make it do anything she wanted' (323). With this knowledge, Nell feels free to leave the protected enclave of neo-Victorians and venture into the larger world. She gains agency from her understanding of the social world as both a construction – rather than a natural given – and as a program that must continually run to reproduce itself – which is therefore subject to change.

As Nell moves within more varied contexts in the social world, she comes to question the representations of the *Primer* itself and its neo-Victorian sensibility that members of the aristocracy are inevitably superior in quality to ordinary citizens. The *Primer* writes, 'She also bought a plain, unmarked saddle so that she could pass for a commoner if need be – though Princess Nell had become so beautiful over the years and had developed such a fine bearing that few people would mistake her for a commoner now, even if she were dressed in rags and walking barefoot' (386).[17] Nell reflects upon this passage, observing, 'Princesses were not genetically different from commoners' (386). Her reflections on this gap between representation and experience allow her to begin to understand how the program of ideology constructs not only the material world, but also the individual subjects within it, with their particular features, skills, attributes, and social standing.

Lola also reflects on the gap between representation and experience.

As I argued above, her struggle for identity and community is related to the gap she perceives between her experiences of friendship with Iz and Jude and her lesbian desire, and the pejorative representations of ghetto people and queers. Unlike Nell, Lola does not gain a sense of efficacy and agency through perceiving this gap. Lola's perception of the gap between representation and experience is also articulated in the novel through her response to information conveyed by the television media. While living in her original neighbourhood, Lola suspects that the television news is 'fudging' (*Random Acts* 18) when it says that everything is fine. She comes to this conclusion based on realizing that her parents regularly minimize problems when they themselves can't face the reality: 'On TV tonight they showed the President meeting with the cabinet. I looked at his face and it looks like Mama's, I don't mean they look anything at all alike. I mean sometimes there just isn't anything there and I think he's on Xanax too' (26). Both Lola and her sister are able to decode the overt statements made on the media and translate them into their 'real' meaning: 'They said no one was killed today. "Nobody important was killed today" Boob said and smiled' (39). However, so long as they remain in their original neighbourhood, Lola believes that the television news will report the events, albeit in a distorted manner. She retains a degree of trust that the media reflect the reality of the material world.

Once the family moves to the new, poor neighbourhood, Lola discovers that the gap between representation and experience is far larger than the 'fudging' that she had assumed. She begins to understand that the media is not simply minimizing the problem in the way that her parents do; instead, she realizes, the media is presenting an edited version of the material world in which some subjects matter (and appear), while others do not (and are absent). Once she lives in the neighbourhoods that are being classified as the 'domestic problem' against which the government must act, she recognizes the extent of the gap: 'We stopped at 137th Street. "Eye that way" Iz said. About ten blocks up there was a lot of black smoke and police cars and fire engines with their lights flashing. There hadn't been anything about more riots on TV so I thought it must have just started. "Going down like this a week now" Iz said' (106). After this experience, Lola learns to distrust the representations on the television and realize that 'there's probably a lot going on that they're not saying anything about' (108). Although Lola does not reflect upon her experiences to theorize about the power structures that inform the media's choices, Womack provides opportunity for his reader to do so. Shortly after seeing the riot on the street, Lola reports on the contents of the evening news: 'On the local TV tonight they didn't say anything about the riots right up the street. They interviewed a radio guy who said the homeless should be killed and the newswoman said really killed? And the radio guy said really killed. Then it showed the dog show' (108).

Lola's relationship to television images provides a critique of post-modern analyses of media culture, such as Baudrillard's, which suggest that people are no longer able to perceive the difference between repre-sentation and reality. Lola's diary indicates that this world of the hyper-real is a world inhabited only by those subjects living in privileged circumstances. When she lived in the middle-class neighbourhood, the television images of the riots did have the quality of hyperreality for Lola. However, once she is physically present, embodied in the moment on the street, the difference between the real and its representation is immediately apparent. The television reports that 'the Army was con-trolling minor disturbances in troubled zones and everything was fine' (163); Lola compares this to her experience of standing on the street corner watching 'about ten dozen Army trucks and humvees and cars coming down 125th Street from the east. They turned onto Broadway and headed north to where the riots were. It looked like an invasion Anne I've never seen anything like it' (116).

It is not so much that the gap between representation and reality dis-appears, as that the standards by which something should be repre-sented change with social context. When Lola's father is alive and her sister is angry about the way Mister Mossbacher is treating him, she announces 'Daddy should hit him on the head with a shovel and bash his brains in' (96). Lola responds, 'It's not cartoons' (96), and explains that their father must tolerate the abusive behaviour because he needs to keep his job. At this point, Lola is still trying to connect with a larger social community and therefore respects its rules of conduct, the differ-ence between reality and cartoons. By the end of the novel, Lola has chosen to 'bash his brains in' because she no longer feels connected to the social community in which such behaviour is inappropriate. Lola still perceives the difference between real life and cartoons, but her con-struction of the social world has changed such that violence is the only appropriate response to Mister Mossbacher. A new context produces both a new body-subject and a new social world as that world is per-ceived/constructed by the subject.

Lola lacks access to a larger perspective that would allow her to per-

ceive the inequitable power distribution that constructs the social world
and its values. In part, Lola is unable to grasp this larger view because of
her age, twelve. Additionally, however, Lola's experience is limited to
what she is able to see in the two social contexts in which she has lived.
Although she clearly perceives the differences between them, these
experiences are not sufficient to allow her to deconstruct the values of
either arrangement. The *Primer* is what allows Nell to learn about things
outside of her day-to-day experience and thus gain a critical perspective
on her experiences. At the same time, the gap between her experience
and the *Primer* allows her to gain a critical perspective on the *Primer*'s
advice. The space for agency and resistance comes from this doubling of
perspective; books are what allow us to step outside the confines of our
material existence, and see our social arrangements as contingent and
cultural choices rather than as necessary and natural givens. Nell is able
to gain agency through her reading because the *Primer* responds and
speaks to her social context. In contrast, Lola is frustrated by the
requirement to read *Silas Marner* and *Tess of the D'Urbervilles* precisely
because they have so little to do with her social context, an irony
Womack surely intends, and cannot help her to negotiate her moral cri-
ses. Lola longs for books that will speak more clearly to her social situation, telling Anne, 'I wish we had other books to read in school. I read
*Life Among the Savages* by Shirley Jackson again tonight. I've read it a
dozen times before, I love it so much' (*Random Acts* 16).

Lola cannot move from education to intelligence because there is no
one to help her mediate the difference between the book and the
world. Given that Lola's daily life requires her to develop skill in avoid-
ing situations in which she may be raped, she finds it difficult to find relevance in Tess's moral crisis about Angel: 'the book's even more boring
than *Silas Marner* and a whole lot more annoying because Tess is such a
wimp so far. I knew Angel would act like he did right from the start he's
such a loser. She should have just gone somewhere off by herself where
no one knew her and start over again I think but I guess it would have
been hard' (115). Lola's experience with reading literature suggests
that, in the final analysis, books are not sufficient on their own to
develop critical consciousness.[18] It is important to have a community of
readers to discuss the representations in books, reflect on them, and
challenge them if there is a gap between their representations and the
readers' experiences. Lola's experiences suggest that if the technology
of reading and writing is to be put to use as a tool to resist dominant
ideology, a collective effort is required.

Both *Random Acts of Senseless Violence* and *The Diamond Age* articulate the integral connection between the self and the social, the reciprocal relationship of construction between them. My reading of these two texts has focused on the ways in which the social constructs the self for Nell and Lola; however, the reciprocity of the connection between self and social suggests that our selves can act on the construction of social reality, as well. Mark Freeman asks, 'If in fact both lives and the stories people tell about them are "socially constructed" and if more generally one cannot ever really step beyond the discursive order inherent in one's own culture, how does one ever manage to go on to do something new and different? How does one ever manage to become conscious enough of the discursive order of one's culture to make transgression and critique possible?' (23). He suggests that writing autobiography, constructing narratives of interpretation and understanding about the self, is the tool of this transformation. He argues that through such writing, one become conscious of the degree to and way in which the self has been externally constructed; this realization allows one to perceive and use those spaces for freedom that do exist.

I would add that the practice of reading and interpreting fiction can function in a similar way. Through constructing readings of a text and coming to an understanding of why the characters become the people they do and make the choices they make, the reader is led to reflect on those external forces which have determined the character. Through identification with such characters, the reader can come to understand the role played by external construction in his or her own life. Every reading is also a rewriting. In our responses to the fiction we read, our evaluations of it as believable or ridiculous, insightful or useless, we make connections between the world of the text and our own social world.

The *Primer*'s automation collapses the role of text and reader: it contains a database of cultural knowledge and invokes specific elements from this database in response to the context it sees around it. In our contemporary world of passive books, it is the agency of the reader that makes these choices, that selects from the available cultural texts and rewrites their tropes according to the experience of the reading self. Writing self is also writing world, in the sense that by writing the story of how one came to be this particular self, one writes one's understanding of the social rules that structure the world in which one circulates.[19] In *Random Acts*, Lola has the sense that she becomes another self, one whom she does not recognize; she attributes the cause of this change to social exclusion, and to Mister Mossbacher, who caused her father's death. Although Womack suggests that Lola's construction of the social world misses the true source of power, we can still understand Lola's construction of herself to be a construction of the world as she now sees it.

In Nell's case, as she reaches the end of the *Primer*, she comes to understand that the *Primer* has been a socially constructed world. In the final chapter of the *Primer*, King Coyote, the creator of the Turing machine castles of the Land Beyond, hands the kingdom over to Nell. Once she has been able to deconstruct the way in which the world has been constructed, she gains the power and the responsibility to 'make new worlds for other people to explore and conquer' (*The Diamond Age* 445). The novel suggests that the knowledge of world construction that Nell has learned moves from the *Primer* to the material world with Nell's role in the Fist[20] uprising at the end of the novel. Nell moves from being a leader, a Princess, in the discursive world to being one in the material world as she and her Mouse Army (the Asian girls) assist in the evacuation of refugees after the Fist attack. The argument of the novel has been that culture forms tribes; Nell is the leader of the new tribe of Asian girls, formed by the unique culture of the *Primer*.

It is important to remember, however, that Nell is able to move from her discursive identity to a material one because she has the support of a community – the Mouse Army – in this move. Her newly articulated self is not simply a private fantasy or psychosis, but an identity which is culturally intelligible to the Mouse Army. As well, I am not suggesting that books alone make the subject, although they have in the very isolated case of the Mouse Army, a community of girls who have had no cultural experiences outside of the *Primer*. The necessity of this community support is evidenced by the fact that Nell is able to use the *Primer* as a tool for social change, while Fiona – who lacks a community of readers to share her story – uses it only for escapism. The parallel between making worlds in the *Primer* and making changes to the material world is located in Nell's insight that each world functions similarly to a Turing machine; each has a defined set of rules that can be used to produce predictable results. Nell's ability to effect change in the material world is predicated on the fact that her insight allows her to understand and manipulate its rules.

Both Nell and Lola move from discursively articulated identities – as the leader Princess Nell and as the violent Crazy Lola – to enacting these identities in the material world. Their ability to move between text and material existence suggests that readers of these novels can also take the knowledge gained from reflecting upon them, and apply this knowledge to the material context of their lives. Additionally, Princess Nell remakes the world through the power of books: her knowledge of nanotechnology and Turing machines is embedded in copies of books that she finds in libraries within the *Primer*. When King Coyote passes on the task of world making to Nell, her first initiatives are to make copies of all these books for all of the girls in the Mouse Army and to begin her autobiography. Nell understands the power of discourse to shape her subjects and works to produce the books that will help her in her task: 'The Land Beyond had vanished, and Princess Nell wanted to make it anew' (*The Diamond Age* 462).

However, as I suggested in my discussion of Lola's alienating experience in reading for her English class, books alone are not enough to shape social subjects and thereby the social world. The representations in the books must be linked to experiences in the real world, or their readers will not perceive them to be relevant. Textual representations do not have to 'match' real-world experience exactly; in fact, the gap between representation and experiences is the space where agency can emerge and change can be theorized. However, the discursive representations must have some link back to material community in order to move beyond the text. A community of readers who can share one's perception of the gap between representation and experience is one of the ways in which this material context can be created. Lola has no one other than Anne with whom to share her perception that she is not 'Crazy Lola' and she therefore fails to resist the social structures that would interpellate her as abject. Community endorsement is required for one's self-representations to have any efficacy, a perspective that belies the liberal humanist faith in individual achievement.

Nell is successful at discursively articulating a self which eventually translates into giving her more power in the material world than she had originally. Lola, on the other hand, feels that her true self is being stripped away from her and the new self she discursively articulates is a product of outside forces, beyond her control. Why do these two girls have such different experiences? The answer lies in the fact that the social community acknowledges Nell's representation of herself as Princess Nell, while Lola's representations of herself are not socially validated. In short, Nell is able to become a body that matters in the sense of materializes in the material world, while Lola cannot. The contrast between these texts mirrors Judith Butler's insight that social representations need to compel communal belief in order to be effective.[21] Self-representations can succeed at socially constituting the subject as represented only if they compel the belief of the community. In other words, social existence is always interpersonal.

Nell is acknowledged as Princess Nell when Queen Victoria II addresses her as an equal. Victoria's ambassador notices that Nell is somewhat taken aback when she is addressed as Your Majesty, and reflects, 'until she had been recognized in this fashion by Victoria, she had never fully realized her position' (*The Diamond Age* 492). The communal belief – that of both her own subjects and the leaders of other tribes – in Nell's authority to be Princess Nell means that she can successfully become the self she describes. Lola's attempts to articulate herself, on the other hand, are characterized by her efforts to deny how others represent her: she is not a queer or, if she is, it is not a bad thing; she is not a ghetto person; she is not crazy. Lola is never able to move beyond this negative rejection of her place in the discourse of others to a positive articulation of her own construction of self.

Both Nell and Lola are examples of subjects who use the power of discourse for social critique. Sidonie Smith, writing on women's autobiography, has argued that writing an embodied autobiography allows the writer to '[come] to an awareness of the relationship of her specific body to the cultural "body" and to the body politic' (131) and that in reflecting on this relationship to write about it, the writer can articulate cultural critique. Both Lola and Nell bring this critical consciousness to their reading/writing practices, Lola in her rejection of the pejorative construction of queer, and Nell in her questioning of Victorian values and aristocratic privilege. Smith, like Freeman, theorizes about the practice of autobiography and relates this power to resist interpellation specifically to the practice of writing. I suggest that the practices of writing and reading are similar in this regard, and that reading a text critically opens a similar space in which one can reflect upon cultural interpellations and perform cultural critique.

My reading of these two novels suggests some of the problems with using fiction as a tool for social critique and social change. Just as the *Primer* cannot ensure that each reader will be produced as the same subject through her engagement with its contents, no novel can predetermine the path that the reader will take through it. In attempting to understand the ideological effects of a text, or in attempting to intervene in the dominant construction of ideology through our textual representations, we must pay attention both to the text itself and to the context within which it is read. A text will not perform the same ideological work for all readers. Just as the body must be conceived of as a

Möbius strip, a blending between inner and outer, self and social, in which it is impossible to determine where one ends and the other begins, so, too, must the practice of reading be understood as an inextricable blending of the ideological constructions of the text and the personal subjectivity of the reader.

Nell's ability to remake herself into a subject able to rise above her disadvantaged birth suggests that fiction can function as a technology to remake our selves. Through remaking our selves, we can work to remake the social structure of the world, to challenge the hegemonic ideological configuration. However, Lola's failure to remake herself as the discursive self narrated in her diary suggests the limitations of such a project. The success that reverse discourses – either discursive or performative – will have in challenging the dominant discourse is contingent upon their ability to compel belief in a wider community of social subjects. Achieving this status of authority is essential to representing a self which can be successful in challenging dominant ideology. This need to speak from a position of authority in order to be successful in one's attempts to intervene in the ideological construction of the social might, at first glance, appear to challenge the idea that popular culture texts can effectively intervene in ideological debate.

Judith Butler considers this issue of speaking from authority in her analysis of the attempts of victims of hate-speech to answer their detractors by recovering and reusing the words spoken in hatred. Butler asks: 'If the performative must compel collective recognition in order to work, must it compel only those kinds of recognition that are *already* institutionalised, or can it also compel a critical perspective on existing institutions?' (*Excitable Speech* 158). Her answer is that subjects who speak from a position that is not authorized can compel a critical perspective on existing institutions, not by using the words of hatred, but by using the authorized speech that those in authority use to describe themselves. The articulation of terms such as 'freedom,' 'justice,' or 'natural right' by those subjects demarcated as excluded from them opens up to public scrutiny and debate the categories of authorization themselves. Butler's strategy is similar to Rosemary Hennessy's description of ideological struggle, which she characterizes as the task of 'sift[ing] through [the current hegemonic ideology's] elements and see[ing] which ones can serve to maintain the interests of a new ruling group ... it is both a process of contesting the articulating principle within a hegemonic formation and a process of disarticulating discourses from one frame of intelligibility in order to rearticulate them in another' (76). Hennessy's notion of the process of disarticulating and rearticulating discursive elements addresses the issue of socio-historical context in attempts to rewrite the self.

SF texts can function similarly. Their relationship to 'real' science is something akin to speaking without authority. However, I would argue that, in so speaking, they offer a critical perspective on the discourse of authorized speakers. As I have argued in this book, SF texts can offer insight into the social consequences of new technologies such as genetic engineering and virtual reality. SF texts appropriate the authority of the scientific speaker to comment on the social implications of these technologies, and attempt to intervene in the types of subjectivities that are forming through human interactions with technology. Science fiction restores – albeit imaginatively – a construction of material, social reality to the technological context. In so doing, it offers us ways to engage with these technologies imaginatively and to choose the types of selves and the type of social that we will allow such technologies to create. Thus, the kinds of posthumanism that appear in SF texts function as both potential models for and current critiques of the ways in which technology and culture are producing a new model of human identity. Reading and writing are tools of the self that participate in this construction of the posthuman just as much as are genetic engineering and virtual reality.

In my reading of *The Diamond Age*, I demonstrated that what the *Primer* shows us is that one's path through a text depends upon one's previous constructions of subjectivity and one's material conditions. SF can reshape subjectivities, but neither the author nor the critic has any guarantee that it will do so in desirable or intended ways. The writing of reverse discourses is a tool for social and subjective change, but – like any other technology – its effects cannot be known in advance, or, ultimately, controlled by the original producer. There are two important conclusions to draw from this discussion of reading and writing as a technology of the self. The first is that science fiction participates as a material technology in the construction of the posthuman; the second is that rewriting the self through reverse discourse is only successful in the context of community belief. The discursive and the material are equally important sites for analysis and for intervention, and an effective model of posthumanism must be one of an engaged, social subject, not an isolated individual.

# Conclusion: Toward an Ethical Posthumanism

The 'end of the human' need not necessarily entail a choice between 'imper-sonal deterministic technologized posthumanism' and 'organic unmediated autonomous "natural" subjectivity,' but may involve modes of post/humanity in which tools and environments are vehicles of, rather than impediments to, the formation of embodied identity.

Elaine Graham, *Representations of the Post/Human*

Two central arguments have structured this book: first that discursive struggles over representation are also political struggles about valid sub-ject positions, and second that we are currently in a moment of defining a new human subject, a posthuman subject, whose features and implica-tions will be intrinsically bound up with the assumptions about identity and embodiment that inform it. Many thinkers seem convinced that the next stage of human identity is just around the corner. This notion that we are living through the last generation of this version of humanity is perhaps most famously argued for by Vernor Vinge in his essay 'What Is the Singularity,' first presented at the VISION-21 Symposium in 1993.[1] In this essay, Vinge argues that 'we are on the edge of change comparable to the rise of human life on earth,' a change that he says will happen not before 2005 and not later than 2030. Vinge identifies computer technol-ogy as the source of this inevitable change based on the development of machines with greater than human intelligence. Such a breakthrough, which he calls the Singularity, will mark 'a point where our old models must be discarded and a new reality rules.' Vinge worries about what the singularity might mean for humans, since super-intelligent machines would clearly no longer be our tools and might in fact be hostile to us.

Vinge makes a distinction in his paper between two possible models of posthuman intelligence after the Singularity, AI or artificial intelligence, with which we are familiar, and IA or intelligence amplification, a kind of super-intelligence he posits might be created by a complete human-computer interface. In either case, Vinge suggests that the posthumans will have little concern for the ordinary humans left in their wake and hence we have reason to be frightened of the Singularity. Despite these concerns, Vinge concludes that the arrival of the Singularity is inevitable for essentially commercial reasons: implicit in his argument about the development of a computer super-intelligence is capitalism and human greed. He argues that even if we know that we should not pursue research into super-computers, or that we should try to put constraints on such research, since the 'unfettered' versions will be 'inferior' to those produced by the risky research, 'human competition would favor the development of those more dangerous models.' Vinge imagines differently embodied humans and perhaps super-intelligent ones, but sees no options other than master race and subservient race for the relations between posthumans and humans.

I would like to take Vinge quite literally at his word that 'old models must be discarded and a new reality rules.' However, instead of just discarding our old models of human capacities or morphology, I argue that we need to discard our assumptions about human nature and the relationship between identity and embodiment, assumptions that I have argued in earlier chapters are all too often still rooted in Cartesian dualism and liberal humanism. As Foucault makes clear in *The Order of Things*, the idea of man as the cogito has a specific historical beginning, and perhaps we are now reaching the moment where that vision/version of humanity has a specific historical end. Struggles over posthuman identity are thus more than struggles over technology and the ethics of various technological ways of modifying what it means to be human. Rather, they are also and more importantly political. The category of the human has historically been used in exclusive and oppressive ways, and the category of the posthuman entails similar risks. However, despite these risks, I believe that it is important to maintain this category and struggle for an ethical and inclusive model of the posthuman because, like the category of human before it, the posthuman has achieved a status such that we cannot ignore the concept, cognizant though we might be of its dangers.

**Models of Posthumanism**

Perhaps the most famous science fiction vision of the posthuman future is Bruce Sterling's *Schismatrix*. Sterling creates a future which expresses many of the motifs and anxieties that have organized the discussion of the posthuman. In this future, the society is split between two factions of space-going humans who have chosen alternative paths for modifying human identity and extending human life: the Shapers, who use genetic techniques, and the Mechanists, who use computer interface and mechanical prosthesis. In the background of their struggle and almost unseen in the narrative itself are those 'primitive' humans who remained on earth without modification of any kind, subject to the Interdict against further technological development because technology was deemed to be tearing the human race apart. The main story in the novel concerns a struggle over what it means to be human, a struggle connected in the narrative to both the different ways, Mechanist and Shaper, of changing the human body, but also the concomitant ways in which human social organization is changed. A faction in this struggle, the Preservationists, tries to defend human cultural purity and recover the great artistic works of the past that are languishing in neglect. The fact that human bodies no longer seem to hold together the dispersed human colonies – that humans no longer have even the basis of common embodiment – is one of the reasons for this turn to culture as what is 'essential' and unique about humanity.

Sterling's posthuman factions eventually learn, however, that the search for a static and unchanging essence of the human is a futile one. The main character, Lindsay, born a Shaper, is in his youth an advocate of Preservationism, insisting upon the purity of the mechanically un-altered human. Over the course of his long life he embraces many different identities and philosophies, and ends by becoming a hybrid somewhere between Mechanist and Shaper, augmenting his Shaper training and chemistry with a prosthetic arm. Lindsay's change does not come suddenly, but rather gradually and always through his ability to adapt. This adaptation does not come without a cost, but it always seems worth the cost to Lindsay because the alternative is dying; life, survival, remains the most important value.

The Schismatrix posthuman cannot become attached to any particular mode of embodiment or any ideology. Flexibility and change are Lindsay's most important – and only stable – skills. Lindsay adjusts his

loyalties and lifestyle as he moves from location to location, always finding a way to thrive through the collapse of governments. Increasingly,
he and those like him turn toward a long-term view, realizing that life,
the species, can persist, but ideologies and social organizations will not.
It is only those who are able to adapt to change, to develop a new dream
when the old one collapses, who are able to persist. The novel, however,
does not criticize Lindsay for this ideological flexibility or make it seem
as if he has lost something crucial in adjusting his old values and losing
what his earlier self would have defined as his 'humanity.' Even the divisions of Mechanists and Shapers shift into a new view that 'the categories are breaking up. No faction can claim the one true destiny for
mankind. Mankind no longer exists' (183). Rather than struggle for the
official vision of posthuman identity, the new generation argues that life
moves in clades, defining a clade as 'a daughter species, a related
descendant' (183). A new version of Lindsay's Preservationists even
arises, the Neotenic Cultural Republic, which refuses membership to
anyone over the age of sixty. Other visions of the posthuman future arise
to compete, but Lindsay embraces the philosophy of clades. He realizes
that just as no ideology persists so, too, no version of posthumanism will
persist. He calls clades 'hopeful monsters that rendered their ancestors
obsolete' but also realizes that 'denying change meant denying life'
(274).

Sterling's vision of the posthuman future is thus ultimately not any
particular vision at all. He does not favour Mechanist visions over
Shaper ones – or vice versa – but he does suggest throughout the reciprocal relationship between how posthumanism is embodied and what
sorts of values and culture said posthumanism will nurture. Sterling also
suggests some of the problems with a vision of posthumanism as immortality: the financial problems of colonies that struggle to support an
aging and technologically dependent population, the clash between
aged and young citizens' world views, and the struggle to maintain interest in life across many decades (near the end of the novel, for example,
Lindsay relies on a drug called Green Rapture to alleviate boredom and
prevent him from falling into a sort of behavourial inertia). At the conclusion of the novel, Lindsay chooses fusion with an alien entity that is
able to survive in any environment. Lindsay and this alien plan to see
the heat death of the universe and then 'see what happens next!' (287).
What is valuable in Sterling's version of posthuman identity is thus not
any particular embodied form or any ideological arrangement of society. Instead, openness to change and newness, to becoming other and

giving up on old categories when they no longer serve rather than defending them against inevitable change, is the mark of posthumanism. The Sterling posthumanist recognizes that human was only ever a temporary category in the first place.

It is no coincidence that those factions in the novel who cling to the idea of a pure human past, including Lindsay's own Preservationists at the beginning of the novel, are also associated with a conservative – in both senses of the word – cultural tradition. The young Lindsay is known for translating the works of Shakespeare into 'modern English.' Sterling ironically suggests that those who choose to freeze culture in such a way are as guilty of deadening culture as those who try to fix humanity at a certain biological and technological moment are guilty of 'denying life.' Lindsay, the defender of human culture, doesn't recognize as plagiarized the tale of 'a group of pirates in the Trojan asteroids [who] have kidnapped a Shaper woman' (51), suggesting that those who seek to preserve the past intact do not necessarily understand it. The brief glimpse we get of earth under the Interdict, a vision of cookie-cutter cities and fear of change, suggests the futility of attempts to fix life culturally, biologically, or technologically. Even the perfect transmission of the original is changed into something dead and mechanical as circumstances change around it. It is not accidental that it is the works of Shakespeare that Lindsay chooses to preserve as 'pure' human heritage. The desire to preserve the 'universal values of humanity' in the cultural tradition that has canonized Shakespeare obscures an elitist and exclusionary definition of 'universal human values' in precisely the same way that some versions of posthumanism hide an elitist and exclusionary vision of 'universal human identity' that is to be translated into a common – and supposedly neutral – postmodern subject.

In Lindsay's transformation from a Preservationist to a continually evolving posthuman, the novel suggests that it is impossible and even detrimental to hold on to singular and restrictive visions of 'true' humanity. I'd like to now turn to what I think is wrong with some of the current visions of the posthuman future that do not have the inclusiveness of Sterling's vision or his attentiveness to the relationship between embodiment and subjectivity. Donna Haraway's 'Cyborg Manifesto' with its celebration of cyborgs as a possible 'way out of the maze of dualisms in which we have explained our bodies and our tools to ourselves' (181) has been very influential on visions of the posthuman future. However, as Vivian Sobchack notes, many of these dreams of transcending the limitations of current human embodiment have been constructed via a set of values that are quite different from what Haraway seems to have had in mind. In her article 'Beating the Meat,' Sobchack begins with the observation that following amputative surgery and fitting with a mechanical prosthesis, 'I have come to learn that it's ridiculous (if not positively retrograde) to accept myself "as I am." I have found I can "make myself over," reinvent myself as a "harder" and, perhaps, even "younger" body' (208). Sobchack goes on to explain that this opening is ironic, but that the reader's willingness to accept it as straight points to a problem in our current relationship to material embodiment, a desire to transcend the flesh which really means a desire to transcend mortality. She argues that it is important to resist the seductions of the cyborg body, that infinitely malleable and indestructible tool, and to retain an awareness of our embodied mortality. The performance artist Stelarc, who argues that 'an ergonomic approach is no longer meaningful' (Atzori and Woolford 197) and that we should change the body to suit our environments rather than vice versa, is one example of the former sort of posthumanism.

Sobchack's article points to the reality that struggles over posthuman identity are taking place in the material world as well as within the pages of science fiction, and that these struggles are being fought over material bodies and subjectivities as well as within discursive spaces. To stress the importance of struggling over the values that inform posthuman identity and technologies of body modification that enable the transition to posthumanism, I want to briefly consider a group of humans who are currently working to become posthuman. The assumptions about identity and values that inform this group point to the ethical danger I believe exists in visions of posthumanism that see the body as irrelevant. These people call themselves transhumans (pointing to their understanding of themselves as intermediate subjects occupying a place between 'mundane' humans and the coming posthumans) or Extropians. 'Extropian' is a name based on its root 'extropy,' which is defined in the 'The Extropian Principles' as 'the extent of a system's intelligence, information, order, vitality, and capacity for improvement.'[2] Extropians see themselves as actively working to increase the extropy of the human organism because they 'see humanity as a transitory stage in the evolutionary development of intelligence.' They work to achieve this goal through modification of their minds and bodies using the technologies of mind-uploading, nanotechnology, neuroscience, robotics, smart drugs, cognitive science, and genetics.[3] Extropian goals and philosophy are stated succinctly by their belief in BEST DO IT SO: Boundless Expansion, Self-Transformation, Dynamic Optimism, Intelligent Technology, and Spontaneous Order.[4]

Max More, one of the leading figures within the Extropian movement, has recorded the seven Extropian principles in a document named 'The Extropian Principles Version 3.0: A Transhumanist Declaration.' I quote this title in full because I think it points to two interesting characteristics of Extropian culture. The first is the degree to which Extropians have refigured themselves in the image of computers; the document is a new release of the 'software' (version 3.0), the program that will create the posthuman subject. As the examination of cyberpunk showed, one of the dangers of constructing our selves in the image of computers is that we can come to see others as inferior or obsolete equipment, as old versions that must be removed to make room for the new. Second, More's title points to a heritage of liberal humanism in its use of the word 'declaration,' which calls to mind other historical declarations of rights. My reading of Extropian discourse is that their style of posthumanism is in fact a covert return to a simplified vision of liberal humanism.

The first principle is Perpetual Progress, which More describes as a rejection of 'traditional, biological, genetic, and intellectual constraints on our progress and possibility.' This leads to the second principle of Self-Transformation, a program of personal improvement using body- and mind-modification technologies as discussed above. Through self-transformation, the Extropian seeks to rise above the 'animalistic urges and emotions' that evolution has inconveniently 'left us with.' The third principle of Practical Optimism points to the heritage of liberal humanism in More's argument that Extropians 'take personal responsibility by taking charge and creating the conditions for success.' Like the self-made men of liberal humanism, Extropians efface the operation of social structures to position social subjects in different relations of power. This omission is particularly glaring in the context of the next principle, Intelligent Technology, which again highlights the importance of body modification to achieving posthuman status. The Extropians offer no comment on the effect that relative access to such technology will have in determining who will become appropriately posthuman. Open Society, the next principle, again connects Extropianism to liberal humanism in its emphasis on the freedom of the individual as providing the best ground for a stable social order. The sixth principle, Self-Direction, reveals the Extropian desire for minimal government, another liberal humanism link; More explains that this principle emerges from the Extropian belief that 'taking charge of ourselves requires us to choose from among our competing desires and subpersonalities.' Extropians argue that government sanctions against technology and self-experimentation restrict their freedom to choose. The Extropian focus on government regulation as that which unjustly restricts their ability to experiment upon themselves in pursuit of their goals again suggests their incomplete vision of social reality; material obstacles or the absence of meaningful options from which to choose are not mentioned. The final principle, Rational Thinking, is another link between Extropian philosophy and liberal humanism. Both ideologies posit 'man's agency'[5] at their centre, reducing the rest of the world to material object, available to be worked upon by 'his' active will.

This final value, Rational Thinking, suggests a connection between Extropian ideas and Cartesian mind/body dualism. By separating mind from body, Cartesian dualism constructs a subject that is based on the repudiation of those parts of subjectivity that may be associated with the body. Liberal humanism performs a similar move of producing abjection in its insistence upon individual freedom without any reference to the social structures which interpellate individual subjects with different degrees of freedom. All are discourses which seek to separate self from the rest of the world. A frequent motif in Extropian literature is the modification of the body so that it can survive in outer space. As *The Cyborg Handbook* makes clear, the term 'cyborg' was coined to describe the product of research into this very topic.[6] The Extropians, then, are among the children of the cyborg, but – as Sobchack observes regarding her own love of her prosthesis – they are hardly what Haraway had in mind.

The Extropian posthumanist philosophy has its roots in both liberal humanism and Cartesian mind/body dualism, and is guilty of the same retreat from material specificity into abstraction that characterizes these discourses from which it grows. Extropians themselves do not acknowledge the influence of liberal humanism, claiming instead that what distinguishes them from liberals is their refusal to insist upon equality of outcome. The Extropian FAQ, frequently asked questions,[7] include a question about the 'typical extropians' attitudes towards women, minority racial groups, and people of nonstandard sexual practices.' The response argues that while Extropians cherish diversity and welcome novelty, they also believe that 'selection must be allowed to take place for progress to occur' and that 'enforced equality of outcome leads to stagnation and stasis.'[8] Given their focus on individual freedom and their denial that social structures have anything to do with producing social subjects in particular relations of power, one wonders what the Extropians envision to be the agency effecting this selection. Additionally, although the Extropians do not include any self-representations which would mark their specific embodiment, the very form of the question posed – 'attitudes towards women, minority racial groups, and people of nonstandard sexual practices' – suggests both that we can understand Extropians to be primarily male, white, and heterosexual, and that they perceive these embodiments to be neutral or unmarked, unlike the 'other' bodies posited in the question, the ones toward which Extropians can have 'typical attitudes.'

Extropians argue for a form of social Darwinism that is rooted in conscious control and direction of 'evolution' rather than in slower natural selection. Given their insistence upon individual rights and the freedom to experiment upon one's body, they do not believe that they are arguing for an exclusionary politics. Their refusal to attend to the vagaries of material conditions means that they operate in a fantasy space 'as if' everyone equally had access to the technologies they advocate. The Extropian response to the question of whether or not they are an elitist organization is: 'if the Extropians are an elite, they are an elite which everyone is invited to join, and the only barriers to membership are those imposed by your own force of habit and whatever tendency you may have to think self-deprecatingly, in terms of insurmountable limitations rather than possibilities for development.'[9] In this move, the Extropians both blame the victim for his or her own failure to become an appropriate posthuman subject – 'tendency you may have to think self-deprecatingly' – and deny the relevance of material conditions – there are no barriers except mental ones. Like liberal humanists, Extropians are guilty of abstracting a universal human nature from specific, material embodied subjects; in both cases, the abstraction is used to ground an ideology of individualism that refuses to acknowledge the political consequences of social institutions and practices that interpellate subjects differently.

Extropians participate in the fantasy that Mark Dery has labelled 'escape velocity,' the confidence that a certain level of technological development will enable humans to escape the material consequences of life on earth as embodied beings. Donna Haraway has called this a deadly fantasy,[10] one that refuses to acknowledge that the earth is finite and we are mortal, and that we must learn to deal ethically and responsibly with one another and with the planet and its species if we hope to survive. Like the participants in the fantasy of escape into cyberspace, Extropians lack an ethics; they are not engaged with the present material world because they are too busy planning for the 'next' one.

The sinister conclusion that may ultimately be reached through such abstraction is accurately reflected in an interview exchange between Mark Dery and Hans Moravec. Moravec, with his plan for uploading human consciousness to computers, is one of the legendary figures within the Extropian movement. Dery asked Moravec if he had any concerns about the fate of 'those on the lowestmost rungs of the socioeconomic ladder' in his projected future of uploaded consciousnesses and robotic bodies. Dery reports the resulting exchange as follows:

> Moravec: It doesn't matter what people do because they're going to be left behind, like the second stage of a rocket. Unhappy lives, horrible deaths, and failed projects have been part of the history of life on Earth ever since there was life; what really matters in the long run is what's left over. Does it really matter to you today that the tyrannosaur line of that species failed?
> Dery: Well, I wouldn't create a homology between failed reptilian strains and those on the lowermost rungs of the socioeconomic ladder.
> Moravec: But I would. (307)

This, then, is the ultimate risk of posthumanism as it is currently configured: we can conceive of ourselves as having the same relationship to those body-subjects who do not matter in the current ideological configuration as we do to dinosaurs. This type of posthumanism so distances the subject from his embodied life that he feels that the 'long run' perspective of millions of years of evolution is the appropriate model upon which to base his relationship to other subjects in the contemporary world.

This model of posthuman is limiting because of the degree to which assumptions taken from the Cartesian view of the subject as disembodied mind and taken from the liberal humanist view of the subject as autonomous owner of self who owes nothing to society continue to inform posthumanist thought. Although Moravec is perhaps the best-known advocate of this model of posthumanism through his goal of uploading minds, his is not the only version that betrays its investment in this sort of model of the subject and the relationship between this subject and society. Ray Kurzweill predicts in *The Age of Spiritual Machines* that we will soon be superseded by computer intelligence, suggesting that the triumph of computers is inevitable because 'the advantages of computer intelligence in terms of speed, accuracy, and capacity will be clear. The advantages of human intelligence, on the other hand, will become increasingly difficult to distinguish' (4). Kurzweill suggests that it is a good thing such machines will take over, an assumption similar to that which informs Banks's Culture, since their superiority will improve life for humans as well. The unspoken assumption here is that intelligence or rationality is the most important aspect of being, an assumption whose negative consequences for social community were pointed out by Ullman's memoir. Gregory Paul and Earl Cox, in *Beyond Humanism*, also argue for a posthuman future based on the superiority of machine intelligence, but their model is an example of intelligence amplification rather than artificial intelligence. They imagine us merging with super-intelligent robots and travelling beyond our planet. Each of these models, like the Extropians, focuses on individual transcendence and ignores social community and the fragility of our material, embodied lives.

One of the things that I find most striking about the Extropians is the evident influence of science fiction on their practices and beliefs. Web sites maintained by self-defined Extropians and by *Extropy: The Journal of Transhuman Thought*[11] include lists of books that neophytes should read if they are interested in discovering more about Extropian ideas. These lists include studies of the most recent developments in fields such as neuroscience and robotics, texts of semi-scientific posthuman imagining such as Hans Moravec's *Mind Children*, and texts by science fiction authors whose future worlds embody the vision of posthumanism that the Extropians are working toward.[12]

By pointing out the existence of both genres on their reading lists, I am not suggesting that the Extropians are unable to distinguish between fictional and non-fictional representations. Instead, what I find intriguing is the relationship between visions articulated in the science fiction and real experiments that Extropians either do perform on themselves or else wish they could perform if not prevented by government regulation or a lack of appropriate funding (or technology). Extropians seem to have fallen into the same trap that Anne Balsamo suggests awaits plastic surgery patients during consultations. As the patients watch their features morph on the screen through the medium of visualizing software, they come to believe that identical results can be produced on their flesh. As Balsamo argues, such a belief is misleading because 'how those incised tissues heal is a very idiosyncratic matter – a matter of the irreducible distinctiveness of the material body' (*Technolo-*

*gies of the Gendered Body* 75–7). Extropians and some posthuman theorists, too, forget about the irreducible distinctiveness of the flesh, and seem to believe that if a modification is represented in an SF text, it is only a matter of time – of technological advance – before such a modification can be realized for them. As I will argue in more detail below, the move to abstraction is precisely what prevents the Extropian concept of posthumanism from having an ethical ground.

Rather than rejecting the concept of posthumanism because of these limitations, I want instead to argue for an embodied posthumanism, one that remains focused on a subjectivity embedded in material reality and that seeks to be responsible for the social consequences of the worlds it creates. This posthumanism will struggle to be post to the emphasis on the universal individual as the centre of meaning and worth. My posthumanism endeavours to be a promising monster: to acknowledge difference without hierarchy; to refuse to found its subjectivity on the grounds of repudiation and boundary setting; and to remember that 'the System is not closed; the sacred image of the same is not coming. The world is not full' (Haraway, 'The Promises of Monsters' 327). Such a critical, embodied posthumanism will include a ground for ethics.

### Embodied Posthumanism

Our current anxieties about posthumanism are ultimately anxieties about how the human might be changed by technology. It is important to attend to the representations and values that are used to support or repress the deployment of a technology, and to examine the political ends which they support. To illustrate this point, I want to contrast two visions of the future and the better bodies-subjects that it may hold. The first is from Lee Silver's *Remaking Eden*:

> The final frontier will be the mind and the sense. Alcohol addiction will be eliminated, along with tendencies toward mental disease and antisocial behavior like extreme aggression. Visual and auditory acuity will be enhanced in some to improve artistic potential. And when our understanding of the genetic input into brain development has advanced, reprogeneticists will provide parents with the option of enhancing various cognitive attributes as well. (237)

The second vision of the future comes from N. Katherine Hayles's *How We Became Posthuman*:

If my nightmare is a culture inhabited by posthumans who regard their bodies as fashion accessories rather than the ground of being, my dream is a version of the posthuman that embraces the possibilities of information technologies without being seduced by fantasies of unlimited power and disembodied immortality, that recognizes and celebrates finitude as a condition of human being, and that understands human life is embedded in a material world of great complexity, one on which we depend for our continued survival. (5)

These representations share some common discursive threads. They both envision a future in which technology will change the body in ways that are beneficial to the human subject. They both imagine that technology can expand the field of human influence. However, they differ fundamentally in their attitude toward embodiment ('genetic input' versus 'ground of being') and in the type of human self they see as produced by this technology ('enhanced artistic potential and cognitive attributes' versus 'embedded in a material world of great complexity'). The first vision sees the technologized super-subject as an end in itself, while the second reminds us that this subject must continue to live in a material world of other subjects and ethical responsibilities.

Embodied and ethical posthumanism requires that we embrace a vision of the posthuman that is closer to Hayles's. The difference between these two representations may be articulated in terms of the difference between abstract theory and detailed life. Hayles's *How We Became Posthuman* is a thorough examination of what is left out in the abstractions necessary to scientific theorizing. She attempts to restore some of this mass of detail both through reading science fiction texts in conjunction with the history of scientific developments in cybernetics and information technology, and through telling this history in the form of a narrative that restores the personalities and struggles of those involved at crucial points.[13] The difference between Silver's representation of the future of enhanced bodies and Hayles's cautious optimism about a posthuman subject that remembers it is embedded in the world is precisely this detail of concrete, real-life, situations. It is only by resisting abstraction that we can achieve a critical posthumanism.

The body – abject, material, immanent, and vulnerable – is that which forces us to recall our own limitations and retain an awareness of our connections to the rest of the world and other beings in it. However, the body has also been the site of cultural exclusions and oppressions, a space where culture acts upon and shapes the subject in limiting and distorting ways. The challenge for an ethical, embodied posthumanism, then, is how to retain a notion that the body is integral to subjectivity without falling into the trap of validating an essential and reified body morphology and identity at the same time.

How might we construct a posthumanism that sees embodiment as integral to identity, but which also embraces openness and change as did Sterling's Lindsay? It is essential to retain the image of the body as a Möbius strip, the blending of inner drives and outside conditioning, to represent the full extent of subjectivity and not fall into the falsifications and erasures of abstraction. As Nikolas Rose argues, 'If subjectivity is understood as corporeal – embodied in bodies that are diversified, regulated according to social protocols, and divided by lines of inequality – then the universalized, naturalized, and rationalized subject of moral philosophy appears in a new light: as the erroneous and troublesome outcome of a denigration of all that is bodily in Western thought' (7). The tradition of liberal humanism and the posthuman subject as influenced by this tradition are both guilty of this negation of all that is bodily in their constructions of a universal subject. An embodied and ethical posthumanism cannot embrace this 'erroneous and troublesome' subject, but must instead work to include the full range of human embodiment in its understanding of subjectivity.

To avoid the dangers of abstraction, it is important to give representation to the full range of human embodiment. There is no 'the body': there are only various bodies differentiated by endless permutations of race, class, age, gender, sexual orientation, geographical location, and any other category we use to discipline and value bodies. The refusal to acknowledge this chaotic mass of detail that is embodied existence, the desire to distort it through abstractions to 'the' (dismissed) body and the universal mind, are the flaws of both liberal humanism and Cartesian dualism. Restoring 'body' to Western thought means refusing to allow the construction of a universalized and naturalized (white, male, heterosexual, able, bourgeois) subject, ending the projection of all that is 'body' onto the marked bodies of others (non-white, female, homosexual, disabled, working-class). Insisting that 'body' is part of all subjectivities, that there is no universal and neutral body, means refusing representations of body-altering technologies that refer to this neutral body. Retaining a sense of embodied, material existence and rejecting the idea that there is a single, 'natural,' or best body are necessary for an ethical engagement with body-altering technologies.

Gail Weiss suggests that the intercorporeal relations between a variety of bodies and body images should be the grounds for all ethics. She refuses a theory of ethics that requires us to deny our embodied particularity in order to 'attain the status of impartial moral agents' and suggests instead that ethics necessarily emerge from the particularities of the embodiment of moral agents because moral decisions 'are not reducible to abstract, rational deliberations that take place between one mind and another mind in a phantasmatic intellectual space' (*Body Images* 158). Moral decisions are not simply ideas, but have material effects in the world, material effects on material, embodied moral agents. Weiss argues that if we take seriously that our bodies are our selves and not just an adjunct to a subjectivity rooted in the mind, then we must also take seriously her demand to root our ethics in bodily imperatives. She models her theory of bodily imperatives on Kant's categorical imperatives, suggesting that the bodily needs of others can make demands upon us that invoke a moral system beyond the abstraction of reason. The example she gives to illustrate her theory is Simone de Beauvoir's retreat, during her mother's illness and death, from the existentialist imperative to face the truth. De Beauvoir was willing to lie to her mother about the seriousness of her illness, and Weiss argues that the demand of the mother's body – her need to deny the truth to face her last days – outweighed de Beauvoir's rational ideas about truthfulness.

I take seriously the notion that if our bodies are our selves, then we need an embodied ethics. Further, I agree with Weiss's observation that social oppression based on bodily differences oppresses the body/subject and therefore demands an embodied response.[14] However, it seems to me that her example of the bodily imperative of the dying body is rooted as much in the social relationship of mother and daughter as it is in the demands of the body itself. Weiss, herself, is cautious in describing the beginnings of her system of bodily ethics, and notes some of the risks when she remarks:

> While I do not think, as some care theorists do, that we must heed the call of all those human (or even nonhuman) bodies who need and/or demand our assistance, I do think that developing a sensitivity to the bodily imperatives that issue from different bodies is a necessary starting place for our moral practices. Which bodily imperatives we attend to will depend not upon some abstract teleological framework which places a higher value on some bodies as opposed to others (e.g., human over nonhuman, those I know best over those I know least, my own body versus other people's

bodies), but rather, must always be a function of the bodily context that situates our relations with others. (*Body Images* 163)

While I agree with the sentiments expressed in Weiss's argument – that embodiment should factor into our ethics – I do not find that she provides an adequate starting point for adjudicating among the competing demands of various bodies. There are many difficulties raised by an ethics of bodily imperatives. Does one person's demand for privacy of his or her body outweigh another's demand for sexual access to that body? Does the demand of a fetus for incubation come before a woman's demand to be the sole possessor of her body? Does an animal's demand for life factor into a human's demand for food, or for knowledge gained by medical experimentation?

It seems to me that in attempting to answer the questions above, we return to the realm of 'common sense' or ideology, the 'abstract teleological framework' Weiss hoped to avoid. It may seem 'obvious' that most people would weigh one person's right to refuse sex over a demand of another's body for sexual gratification. However, the rights of the fetus for incubation put us on less secure ground. Does the woman's right to control her own body still apply, given that the fetus faces death, rather than mere sexual frustration, as the consequence of her refusal? Despite Weiss's hopes that we can articulate a bodily ethics without reference to some abstract system which values some bodies over others, it seems impossible to resolve these competing demands without doing so. Is the fetus a person? Does it have the right to gestation? The same questions are raised when we consider the competing demands of human and non-human bodies. An ideological system which values animal bodies less than it does human bodies is invoked to justify meat eating and medical experimentation. In the absence of this abstraction, how could the competing claims between the human and non-human bodies be resolved? Therefore, while I am sympathetic with Weiss's objective in arguing for an embodied ethics, I do not believe that it is possible to formulate any ethics in the absence of ideology. A bodily ethics constituted without reference to 'some abstract teleological framework' is, in fact, as abstract as liberal humanism. In each case, the ideals expressed in the discourse cannot be achieved in practice.

Weiss calls for our ethics to be a function of the bodily context that situates our relations with others; part of this context will be discursive representations that structure the domain of social intelligibility. Although Weiss wants to argue for a bodily ethics that remains cognizant of the particularities of our own and other bodies, she has failed to provide a practice by which we could make ethical choices among bodily demands. There is an unacknowledged universal in Weiss's work, just as there is in liberal humanism, a universal that suggests that there is some common set of values and standards to which we can look. Weiss's bodily ethics installs a universal body at its centre, just as Kant placed universal reason at the centre of his ethical program. Weiss's universal body is evident in the fact that she feels that we can found a bodily ethics outside of ideological struggles over bodily values. Such a move suggests that we can understand 'the' or 'a' body outside of the discursive representations that produce its particular contours; it suggests that there is an outside to struggles over bodies that matter in both meanings of the term, that is, bodies that are deemed ideologically important and bodies allowed to materialize in discourse.

As I have argued throughout this work, I do not believe this to be the case. It is important to keep challenging the range of bodies that matter, so that these bodies will be taken into account when making ethical choices. In our engagements with both genetics and information technology, it is necessary to retain a sense of embodied subjectivity, of real material consequences to our actions and choices. This continued articulation of embodied subjectivity in our representations is, I suggest, the real starting point for an embodied ethics. Rather than trying to find a neutral or innocent ground from which we may articulate our bodily ethics and judge among the competing demands of various bodies, we should instead recognize this move as another attempt to establish what Haraway has called the gaze from no where.[15] Haraway calls knowledge claims made from this gaze from no where 'unlocatable, and so irresponsible' ('Situated Knowledges' 191) because they cannot be called to account for their constructions and investments. My notion of an embodied ethics would require that – rather than try to disavow our dependence upon abstract frameworks which place a higher value on some bodies as opposed to others – we instead acknowledge the positions from which we speak and the social constructions of value that speaking from such positions entails.

Such a strategy for ethics refuses the seduction of constructing one's speaking position as neutral or innocent; it requires that we acknowledge and defend our own ideological investments, our own definitions of bodies that matter. Haraway has called this situated knowledge, a type of knowledge claim that acknowledges that its perspective is always partial, the view from a particular location. She argues that 'partial perspective can be held accountable for both its promising and its destructive monsters ... Feminist objectivity is about limited location and situated knowledge, not about transcendence and splitting of subject and object. In this way we might become answerable for what we learn how to see' ('Situated Knowledges' 190).

It is only by remaining faithful to the material context in which we and other subjects are embedded that we can begin to negotiate a collective bodily ethics, an ethics that acknowledges our own and others' ideological investments in valuing particular bodies over other bodies, and works to make ethical decisions within this web of bodily demands. Bodily ethics cannot remain free from the realm of ideology; instead, it is important to ask the question of who benefits from these choices, and to take responsibility for the consequences of our situated knowledge choices. For example, in an attempt to mediate between the bodily demands of animals and the bodily demands of humans on the issue of laboratory tests using animals, those who argue in favour of the animal bodies must be willing to be accountable for the way in which their position entails the suffering of human bodies from disease, while those who argue in favour of the human bodies must be willing to be accountable for the suffering of animal bodies. Being accountable does not mean that one is personally responsible, as the individual author of the act, but means acknowledging and accepting that such consequences are part of the material context that situates the relation of these bodies. Revealing our ideological investments, articulating a partial perspective or situated knowledge rather than attempting to work outside of ideology and therefore falling into the distortions of abstraction, is the necessary starting point for a bodily ethics.

Donna Haraway's description of science fiction as the domain for the interpenetration of boundaries between problematic selves and unexpected others has been a motif that has guided my study of the intersection of body, text, self, and social. Throughout this book, I have described the ways in which this interpenetration is an aspect of subject formation: we define what we are through reference to what we are not as much as through reference to what we identify with. Again and again, the problematic selves are revealed to be the alterity that is necessarily incorporated into the self, and the unexpected others are the others that are made other by the practice of boundary setting in subject formation. The other is only the pole of intelligibility of self in the current cultural formation, a pole subject to change.

Unexpected others and problematic selves thus produce one another.

SF is the discourse of boundary figures – monstrous and otherwise. Through its representations of figures that push the limits of intelligibility, SF is able to raise questions about what it means to be human, and to suggest definitions that were previously invisible, untold and therefore impossible.[16] Vicky Kirby suggests that the current 'crisis of the subject' might be understood as a crisis of species identity as much as a crisis of subjectivity. She argues that, instead of focusing on how '*the humanist subject* is actually decentred, that the individual cannot secure his or her identity through intention, and that individuation as such is necessarily contingent,' we should think about how '*the subject of humanness* recognizes itself as a unified subject, individuates itself within species-being, and identifies itself as possessing sufficient stability to ground the destabilization of grounds' (151). Kirby suggests that philosophical grounding for the humanist subject is being re-established based on the subject of humanness, the attempts to articulate a stable identity based on differentiation from other species.

In Kirby's view, the boundary between human and non-human (animals or machines) is reproducing the Cartesian mind/body split. The new binary of human/non-human again jettisons the body from subjectivity, making the human coincide with qualities considered to be mind/male by Cartesian dualism. In my view, ethical posthumanism needs to move away from this subject of humanness; the 'post' of post-humanism should not be a post-biological embodiment. The 'post' of posthumanism should be a 'post' to the heritage of humanism, which makes humans the only subjects in a world of objects. An ethical posthumanism must work against this boundary of the human from the non-human, refusing this final ground of abjection. An ethical posthumanism which acknowledges that self is materially connected to the rest of the world, in affinity with its other subjects, is an accountable posthumanism. It is a posthumanism that can embrace multiplicity and partial perspectives, a posthumanism that is not threatened by its others.[17]

Science fiction is an excellent resource for interrogating how we construct the posthuman, and the political ends inherent in various constructions, because its generic conventions provide a space for narrating agency for non-human subjects. It is important to examine the boundary between human and non-human if we wish to theorize a truly embodied subjectivity, one that does not let us project the seemingly negative qualities of the body onto the non-human. Examining and challenging this boundary is also essential to any attempt to articulate a bodily ethics. Classifying bodies as non-human is a well-established justification for condoning their abuse and exploitation. The construction of the posthuman subject, mergers of human with machine and human with animal, is similar to the deployment of a technology. It is neither emancipatory nor oppressive in itself. It can result in the recognition of our affinity with these non-human subjects, but it can also create a reactionary attempt to articulate more forcefully the boundary between human and non-human in an attempt to disavow this affinity.

The values which underlie our technological choices are the most important determinant of their consequences, and this fact is why I have been so insistent upon critiquing the persistence of some damaging assumptions from liberal humanist thought in some versions of posthumanism. With Hayles, I prefer a posthumanism which signals the end not of humanity but of a certain conception of humanity, that 'fraction of humanity who had the wealth, power, and leisure to conceptualize themselves as autonomous beings exercising their will through individual agency and choice' (*How We Became Posthuman* 286). While it seems inevitable that technoscience will change what it means to be human in the near future, this is not necessarily a cause for despair. It is imperative that we develop an ethically responsible model of embodied posthuman subjectivity which enlarges rather than decreases the range of bodies and subjects that matter. Such representations are the path to an ethical, accountable, embodied posthumanism, to being more rather than less human in our next iteration.

# Notes

## Introduction

1 See Donna Haraway, *Modest_Witness@Second_Millennium* 205ff for examples such as varying nutritional advice given to white and non-white pregnant women, varying infant mortality rates between whites and non-whites, sterilization as a birth control practice recommended to poor and non-white women, etc. See also Best and Kellner 149–204 for a discussion of various intersections of capitalism and genetic technology, described by terms such as biopiracy or bioprospecting to mark the ways in which the genetic material of the body itself has become a resource for accumulation by corporations without benefit to the very people whose bodies are so harvested.

2 It is estimated that hermaphrodite births account for 4 per cent of all live births (Nelkin and Lindee 125).

## 1. Gwyneth Jones: The World of the Body and the Body of the World

1 Clearly 'Aleutians' is an odd name for an 'alien' race, since it describes a geographical location on earth. As I will argue in this chapter, Aleutians do not see themselves as radically Other from the people of earth, and they do not often use spoken language among themselves. In order to communicate with humans, they adopt earth languages, primarily English. The Aleutian Islands are the location where the aliens first reveal themselves to humans, leading the humans to use the term Aleutian to refer to the aliens. When the aliens are later asked what their home is called, they reply 'Aleutia.' From the alien point of view, the place does not need a formal (that is, spoken) name unless they are speaking to outsiders. They see the use of the name Aleutia to describe their home as a way of translating from the Common

Tongue, which is not spoken aloud, to English. As an Aleutian character explains it, 'Our articulate languages are extremely fluid and contextual … People change their spoken names, and the names for things, constantly. In the context of the Expedition to Earth, it's natural for us to use Sanskrit or English or whatever seems appropriate' (*Phoenix Café* 142–3).

2 The rape of Johnny is a central moment in the trilogy, and I will discuss this incident in more detail below.

3 I will discuss the significance of human versus Aleutian use of faster-than-light travel in more detail below.

4 The work of Julia Kristeva on abjection is often important to discussions of the role of repudiation in subject formation. I have chosen not to use Kristeva's work in this chapter because I believe that her model of abjection is rooted in an ontological understanding of certain universal categories of the abject. Kristeva's argument addresses instances where specific groups of individuals have historically come to be associated with this transcendent abject category, but in my view she relies too heavily on the notion that the abject as a category is a psychological universal linked to transcultural experiences of death and separation from the mother. The critical thrust of my argument relies on understanding the abject as a *socially produced* category that is always contingent upon cultural time and place. Kristeva's notion that it is self that is rejected through the construction of the socially abject is important to the notion of subject formation that I use in this book. Her notion that it is self that is expelled is based on her understanding of this process in the young infant who has not yet had the chance to develop the distinction between self and world. It seems to me, however, that the larger trajectory of her work is that the abject is an essential category. Her reliance on Freudian psychoanalytic categories and her insistence upon seeing the abject only in terms of a misogynistic repression of the maternal seem to limit Kristeva's usefulness as a theorist for understanding the *political* and *social* effects of repudiation.

5 Kristeva also discusses this realm of the prelinguistic that exceeds the domain of the Father's law and language. However, it seems to me that her usefulness on this topic is limited because she insists on associating this space with what is designated the feminine, as an other without a name, the unnamable otherness that the subject is confronted with via the paternal prohibition of the Oedipal triangle. While what is considered 'the feminine' is one possible source of what is lost in assuming a subject position, it does not seem that this is the only or necessary source of that excess which cannot be expressed in language systems.

6 In her reading of the first novel in the trilogy, *White Queen,* Jenny Wolmark

argues that this narrative style means that all viewpoints are presented as pro-visional and transitory, denying the notion that a dominant hegemonic inter-pretation as 'true' is possible.

7 The angled brackets are used in the text to denote exchange between the Aleutians in what is called Common Tongue, while conventional punctua-tion is used to denote speech exchanged audibly.

8 Throughout the trilogy, the aliens fail to understand the human distinction of sex/gender. In their notion that they are all 'people' they use the mascu-line pronoun to denote the neutral pronoun, as has largely been the prac-tice in human languages. The Aleutians are often insulted when referred to by the pronoun 'she,' as they see this as an attempt to reduce them to a less-than-person subject position.

9 Unlike Haraway's cyborg, the Aleutians are a representation of this crossing of boundaries as something that is 'natural,' an inevitable part of their being-in-the-world. Haraway, in contrast, argues that the cyborg resists being coded as natural; one of its strengths as a metaphor is that it forces one to realize that it is made, not found, as are all objects of human inquiry. In an attempt to reconcile this apparent difference, I would argue that it is our desire to read the body as natural that creates a tendency to see any activity or product of the body as part of the natural. In fact, the aliens exercise a form of bio-technology to construct their commensals – animal machines – that could as easily be read as cultural.

10 It is interesting to note Johnny's disgust and fear of the interpenetration of self and other in terms of the discourse on AIDS that currently circulates. AIDS is about fear of the contamination of the other into the self. In *White Queen,* Johnny is infected with a virus called QV which is related to AIDS in terms of the way discourses about the virus circulate. Johnny is ridiculed about having 'sex with machines,' which is reminiscent of discourses which construct AIDS as a punishment for inappropriate sexual activity. Johnny is required to enact containment protocols during all his bodily contact with either humans or machines, a vivid representation of how fear of contamina-tion works to isolate the afflicted individual.

11 It is also important to point out that Johnny and Braemar have a pattern of her appeasing his violence with the offer of sex. Although Johnny does not make a sexual move in this specific encounter, Braemar is wondering if she should take off her clothes. Braemar's past as an abused wife is extremely important in understanding the way she interacts with Johnny. She con-structs herself as a victim before he 'has a chance to,' because in her estima-tion her victimization is the inevitable end of these scenarios of rage. She also understands that violence and sex as a mix have to do with humiliating

194 Notes to pages 42–6

her, something some men must do to women in order to maintain a positive sense of self. Although Johnny does feel humiliated by the encounter with Clavel, it is important to note that he does so largely based on his construction of how he has felt when he was the aggressor and that the intention to humiliate is absent on Clavel's part. Clavel is as devastated by the incident as Johnny. Without a concept of separate selves and hierarchy, the logic of rape is not possible.

12 See Haraway, 'The Promises of Monsters' 297.

13 This incident is earlier in the sense of a human history chronology, although the passage comes from a later text. This is one example what I believe to be a strength of Jones's work in this trilogy. Like the humans who are characters in the novel, the human readers are required to assess and construct meaning based on information that is always partial and provisional. The novels require the reader to actively renegotiate, revisit, and revise throughout the reading process.

14 The intersection of AIDS fear with previous fears of miscegenation is demonstrated particularly by hysterical notions that AIDS, which does infect primates, came from Africans who had sex with monkeys.

15 The humans test their experimental weapons on the 'daughters' of Traditionalists (male side in the Gender War). These daughters are revealed to be clones produced from the tissue of the parent – usually the father – who are genetically altered to permanently remain little girls. These 'traditionalist young ladies' are used as sex toys by their families. The Traditionalist humans believe that this creation of another life form simply to exploit it for their own gratification is analogous to the Aleutian creation of commensals to perform tasks. The novel displays the radical distance between the Aleutian affection and consideration for the commensals and the brutal treatment of the young ladies, suggesting again the damage that results from the human failure to perceive a continuity between self and other. Even when the other is made from one's own tissue, humans can still treat it as abject object.

16 Other than specifying that reproduction is parthenogenetic, but that the genetic makeup of the offspring is triggered by some chemical event that 'calls' forth one of the three to five million genetic individuals that constitute Aleutian DNA, not much information is provided about Aleutian reproduction in the novels. Even the Aleutians themselves have no idea what triggers a pregnancy, although stress is one of the proposals. In any case, there seems to be some way of ensuring that a single individual is born only once within a generation. It is not clear whether the total population of Aleutia remains constant at all times, that is, if there is a birth for each death as in Marge Piercy's *Woman on the Edge of Time*. In any case, given this scarcity of information regarding reproduction, it is not clear in what way the tissue is modified

in order to allow it to become a fetus. However, the Aleutian ability to mould desired objects from biological tissue, and the repeated motif that humans and Aleutians are 'no different flesh,' suggest that we are to understand that this tissue remains human at the DNA level, although most of the gross phys-ical features of the body become Aleutian.

17 Bella prefers to be referred to by the pronoun 'he,' since this is the human pronoun that the Aleutians understand to mean a full human being. How-ever, her disability and weakness make her appear female to humans. There-fore, the female pronoun is often used to refer to her in interactions with humans or descriptions from a human point of view, while the male pro-noun is used when the description is from an Aleutian point of view.

18 Jones argues in her essay 'Sex: The Brains of Female Hyena Twins' that human culture has also reached a point where sexual reproduction is irrele-vant to social organization because the evolutionary strategy of producing more children is no longer the one best suited to our survival. She explains: 'We can adjust our sexual behaviour, and thereby our social roles. We can't – not yet – alter reproductive function ... But more and more, for selfish indi-vidual human animals both male and female, reproduction is not money. Money is money: is status, territory, resources. According to the science in *The Differences between the Sexes*, we will all adjust our sexual behaviour accord-ingly ... If sexual behaviour and function are malleable, and yet sexual iden-tity, *difference*, remains obstinately intact – which is what the science predicts and what we see happening around us – then we don't have two complemen-tary sexes any more, each safe in its own niche. All there is left is gender: an us and them situation. Two tribes, separated by millennia upon millennia of grievances and bitterness, occupying the same territory and squabbling over the same diminished supply of resources. This is the situation, the riff that I've found and used, and which you'll find explored in my novels *White Queen, North Wind,* and *Phoenix Café*' (106–7).

19 A classic example of this is Arthur C. Clarke's *Childhood's End.* In some ways, *White Queen* is a revisionist writing of the tropes of *Childhood's End.* As well, Octavia Butler's *Xenogenesis* trilogy, which will be discussed in the next chap-ter, also draws on this motif. Many critics of SF, such as Aldiss, Broderick, and Delany, have argued that this recirculation of tropes and motifs between nov-els is a characteristic feature of the genre.

## 2. Octavia Butler: Be(com)ing Human

1 Butler's use of genetic engineering in her trilogy is not an example of 'hard science' SF, that is, SF which deals minutely and as accurately as possible with the details of the science it represents. Butler's use of genetics is usually read

as a narrative trope that permits her to engage with the themes she is interested in exploring about human culture and values. In my reading of her texts, I juxtapose her representations of genetic engineering with contemporary responses to the 'real' science. My rationale for this approach is that many representations of genetic engineering are in fact arguments engaging with the same themes found Butler's fiction rather than neutral descriptions of science. That is, I am interested in interrogating the assumptions about human 'nature' and about appropriate uses of technology that are embedded in the non-fictional representations. My argument is that many non-fictional accounts of genetic engineering (how we should use it, why we should use it, who should use it) are ideological stances about values rather than simply hard scientific fact – are in fact as 'metaphorical' and 'symbolic' – as Butler's work.

2 All quotations are taken from a combined edition of the trilogy published under the title *Xenogenesis* by Guild America Books. No date is given for this edition. The copyright dates for the novels are *Dawn*, 1987, *Adulthood Rites*, 1988, and *Imago*, 1989. I will indicate which novel I am quoting from in the text, but all page references will be to this combined edition.

3 This ability suggests that in Butler's imagined future, all species in the universe? galaxy? share the same genetic code (four bases) that is shared by all living things on earth. This further suggests that Butler's work shares the perspective discussed in chapter 1 that a more socially constructive way of being-in-the-world would be to consider ourselves as part of the same being as all other living things. As Nelkin and Lindee point out, this conclusion is one that logically emerges from genetic information such as the common structure of DNA in all living things and the fact that human DNA and chimpanzee DNA is different by approximately 1 per cent (126). However, rather than focusing on these similarities, the thrust of genetic science has been to look for differences that distinguish humans from other species and that distinguish normal from abnormal humans. Again, this emphasizes that the nature/nurture debate is far from over: the very 'facts' of genetic science cannot escape being socially constructed by culture.

4 This means 'not Oankali' as they were when they encounter humans, since the history of the Oankali suggests that they keep calling themselves Oankali throughout their genetic transformations. The Akjai – examples of the Oankali as they were before they encountered humans – have distinctly different morphologies from the Oankali that interact with humans in the novels.

5 The Oankali technology, like that of Jones's Aleutians, is organic. The ability of the ooloi has been used to alter non-sentient species in ways that allow

them to serve as tools for the Oankali, and to enjoy their role. Their ship is also a living organism, with which they share 'an affinity, but it's biological – a strong, symbiotic relationship. We serve the ship's needs and it serves ours. It would die without us and we would be planetbound without it. For us, that would eventually mean death' (*Dawn* 38). Interestingly, this sense of organic technology, rooted in genetics and the ability of organisms to adjust to their environment, is similar to descriptions of the organic changes it is believed that earth underwent in its move from a nitrogen-rich atmosphere to an oxygenated one. Lynn Margulis and Dorion Sagan write in *Microcosmos: Four Billion Years of Microbial Evolution:* 'In their first two billion years on Earth, prokaryotes continuously transformed the Earth's surface and atmosphere. They invested all of life's essential, miniaturized chemical systems – achievements that so far humanity has not approached. This ancient high *bio*technology led to the development of fermentation, photosynthesis, oxygen breathing, and the removal of nitrogen gas from the air' (17). Cathy Peppers argues that this alternative evolution story, in which we are in symbiotic cooperation with other species like the mitochondria, instead of engaged in a 'war of all against all,' a struggle for survival of the fittest at the expense of all competitors, is an example of the cyborg origin rewritings that are accomplished by the trilogy ('Dialogic Origins' 54).

6  See Brad Evenson, *National Post*, 19 August 1999, A1. The same experiment results were reported in the *Edmonton Journal* by Lisa Krieger ('Nice Mice Just a New Gene Away,' 22 August 1999, B2).

7  See Jeff Barnard, 'Talk about Building a Better Mousetrap; This Gene Experiment Created Smarter Mice,' *Edmonton Journal*, 2 September 1999, A1.

8  See Lisa Belkin, 'It's a Girl/Boy or Whatever You Want,' *Edmonton Journal*, 15 August 1999, F1.

9  For examples see Silver, and Aldridge.

10  For examples see Nelkin and Lindee, Nelkin and Tancredi, Miringoff, Kevles, and Appleyard. Russo and Cove present a quite balanced view that acknowledges benefits that may result from genetic manipulation of food crops and animals, but warns against the dangers of allowing such technology to be applied to humans. Their analysis includes a discussion of the continuum between eugenicists and geneticists in terms of university appointments in Britain, the US, and Germany.

11  There are two aspects to this argument. Wexler's article focuses on the case of Huntington's disease, a disorder whose symptoms do not emerge until middle age. Wexler cites her experiences in studying the disease in a community in Venezuela and the risk that someone who finds out that he or she is to suffer this debilitation will choose to kill himself or herself rather than

let the disease emerge. The more commonly cited moral objection, which Wexler also discusses, is the 'treatment' of abortion for those fetuses found to have a genetic abnormality. Miringoff also analyses this second aspect of the moral implications of genetic screening.

12 See Farquhar, especially 72–86. The counter-discursive image is that of the hyper-fertile coloured and/or poor woman (the 'welfare mothers syndrome'). Both of these images relate to eugenics discourses from the beginning of the century, particularly in America, which emerged, in part, as a response to immigration and fears that 'superior stock' were being out-bred by 'inferior stock.' Farquhar points out other intersections of social values and genetic science emerging in discursive constructions of 'innocent' infertile women and 'non-innocent' infertile women – those who have difficulty conceiving because of previous sexual promiscuity, delayed child-bearing to pursue career or education, or the wish to pursue parenting outside of a heterosexual couple.

13 See also Bordo 76 and Haraway *Modest_Witness@Second_Millennium* 205 for further discussion of the increased social control exercised by medical practitioners over the reproductive freedom of poor and non-white women as compared to that of middle-class, white women.

14 The more recent ability to select the sex of an embryo during IVF – the Microsort procedure discussed in Belkin, 'It's a ~~Girl/Boy~~ or Whatever You Want' *Edmonton Journal*, 15 August 1999, F1 – is currently restricted to couples who already have at least one child and who wish to select for the opposite sex. However, the experience with other IVF technologies suggests that as soon as the procedure and equipment become more widely available, price will be the only criterion required to qualify. Genetic research and its attendant technologies and procedures are money-making businesses, and governments are unlikely to have much ability to regulate their use in the face of stockholders who wish to see their profit, and a global economy in which genetic research firms can be moved to another country to avoid restrictions (see Silver, and Kevles). Those limits that have been successfully enacted have been done through careful assurance that all countries agree to the same standards. For a discussion of the current controls in place see Watson, and Kevles. In the case of Microsort, the fear is, of course, that there will be an overwhelming preference for one sex. Belkin suggests that North Americans indicate a preference to sex-select for female children, and that this may be because women have more control over their reproduction in the West. It should be noted that this is an expressed preference by people who *want* to undergo the procedure, since there are not yet sufficient numbers of people who have used the procedure to report statistically significant

results. Nelkin and Tancredi report sex-selection in preference for male children through the use of abortion in India (65). The ability to choose which pregnancies to continue based on genetic information about the fetus raises an ethical question for feminists and clinicians, both of whom may be stuck between not wanting to advocate abortion for the reason of trying for a 'better baby' and not wanting to see a woman's right to choose an abortion restricted in any way. Cowan looks at this issue in her article 'Genetic Technology and Reproductive Choice.'

15 See Miringoff, Nelkin (alone and also with both Tancredi and Lindee), Greely, Keller.

16 See Miringoff, especially 18–20. For those who doubt that the practice of selective abortion could start to be applied to an increasingly wide circle of examples, consider the 1985 book *Should the Baby Live? The Problem of Handicapped Infants*, by Helga Kuhse and Peter Singer, which argues that some infants with severe disabilities should be killed. Consider also that genetic treatments for 'abnormalities' have been demonstrably increased in scope. For example, human growth hormone is used to treat dwarfism in children to allow normal growth through puberty. In the past, when the supply of the hormone was limited to what could be extracted from cadavers, the scope of treatment was limited to those children who couldn't produce any of the hormone in their own bodies. Now that genetic engineering can provide an unlimited supply of the hormone from bacteria, the scope of treatment has been extended to those children who are considered 'too short' for their age (see Hubbard and Wald 163).

17 If this seems like an attitude that is rooted in paranoia, consider the fact that in the United States physicians can be sued for wrongful life (the disabled individual sues the physician for failing to treat or prevent the disabled birth) and wrongful birth (the parents sue the physician for the same reasons). It doesn't seem like a long step from this perspective to an argument that says that parents who choose to have a disabled baby despite the predictions of medicine should therefore assume the entire financial 'burden' of their choice. In an economy in which we are continually looking for ways to reduce the economic cost of social programs or transfer responsibility for such programs to the individual rather than the state, it seems entirely feasible that 'special' services that can be identified as non-essential (since the 'special births' could have been prevented) will be targeted for reduction or elimination.

18 See Greely for a further discussion of the ramifications of the intersection of medical insurance and employment in the current American health care system. As most employers are also insurers, the risk is that genetically

'afflicted' individuals will be denied employment opportunities because the insurers do not wish to bear the burden of their medical expenses. As Greely points out, the solution is not as simple as straightforwardly legislating against such discrimination, since the likely consequence of this would be that employers would then choose to stop providing medical insurance as an employment benefit. Further, if private health insurers are prevented from discriminating against those with genetic markers of disease, the economic system which currently underpins American health care insurance (profits for stockholders) will collapse. This situation is further complicated by the fact that 'there may be little correlation between positive tests and impaired performance. Yet data from tests are compelling: though a person may have no symptoms, a diagnosed predisposition to a disease can itself be perceived as a kind of abnormality, a disability, a disease' (Nelkin and Tancredi 102). Robert J. Sawyer's SF novel *Frameshift* considers the problem of genetic testing and health insurance.

19 If all of this sounds suspiciously familiar, see Kevles, and Nelkin and Lindee, for a discussion of the continuing rhetoric between eugenicists from the beginning of the century and geneticists at its end.

20 For a discussion of these various representations of genetics in popular sources, see Nelkin and Lindee. For examples of scientists using them without irony, see Kelves and Hood.

21 This is assuming, of course, that such complex social patterns can be reduced to genetic propensities, which is not entirely certain. The very fact that the desire exists to find such singular genetic markers is worrisome enough. As Nelkin and Lindee point out, we still inhabit an age in which we feel a need to define the limits of nature and the beginnings of humankind. Many technologies challenge what it means to be human: machines can think; VR can simulate experience; animal rights suggest humans are not unique; cyberbodies challenge the idea of an organic base to human uniqueness; sociobiology links human culture to primate (42). More than ever, we now rely on the body – as DNA – to define our identity. A further complication of this dilemma is the fact that if we believe that genetic predispositions determine a person's fate, the knowledge of this genetic script may influence how parents and society treat a child, thereby producing the very subject that was predicted by genetics (see further Miringoff 50). If this scenario seems too close to the image of Huxley's *Brave New World* (and hence unrealistic), consider evidence cited by Nelkin and Tancredi regarding the over-diagnosis of hyperactivity in schoolchildren and the state's insistence that such children be treated with Ritalin in order to be allowed into classrooms (120). It is far easier to diagnose and accept biological explanations for problem chil-

dren – which can quickly be fixed by drugs – than to look at more complicated issues like family stress, overcrowded classrooms, etc. – which frequently require economic and social structure solutions.

22  The default tendency will be for the discourse of genetics to be deployed in socially conservative ways. As Nelkin and Lindee argue, 'Charged with cultural meaning as the essence of the person, the gene appears to be a powerful, deterministic, and fundamental entity. And genetic explanations – of gender, race or sexual orientation – construct differences as central to identity, definitive of the self. Such explanations thereby amplify the differences that divide society' (126).

23  For a science fiction example of a society polarized by genetic modification see Nancy Kress's *Beggars* trilogy.

24  Human fear of the Oankali is visceral. When Lilith is first forced to spend time near one, she must force herself to resist her revulsion. The reason for this fear and revulsion, which easily become hatred, is rooted in the fact that Oankali bodies are different from human bodies. The Oankali have tentacle-like sensory organs, which the humans perceive as snakes. Lilith discusses both his Oankali and his human heritage with her son, Akin: '"Human beings fear difference," Lilith had told him once. "Oankali crave difference. Humans persecute their different ones, yet they need them to give themselves definition and status. Oankali seek difference and collect it. They need it to keep themselves from stagnation and overspecialization ... You'll probably find both tendencies surfacing in your own behaviour ... When you feel a conflict, try to go the Oankali way. Embrace difference"' (*Adulthood Rites* 321). Like hierarchy, Butler diagnoses human fear of difference as one of the causes of our social problems. Additionally, Lilith's statement to Akin about humans needing difference to give themselves definition and status points to the role of repudiation in the constitution of the self.

25  This includes behaviour that continues even after the Mars colony option has been offered to them. Even Akin has his doubts: 'They were not killing each other over the Mars decision, but they were killing each other. There always seemed to be reason for Humans to kill each other. He would give them a new world – a hard world that would demand cooperation and intelligence. Without either, it would surely kill them. Could even Mars distract them long enough for them to breed their way out of their Contradiction?' (*Adulthood Rites* 484).

26  For example, see Wexler on the interpretation of results of genetic screening. As the markers may change from group to group, a negative response can mean either that the disease is not present or that the disease is not marked by one of the ways that may be tested. Nelkin and Tancredi discuss

the various interpretative aspects to any biological testing throughout their work. Eric Lander discusses some of the technical issues involved in DNA fingerprinting and the high potential for error in the procedure.

27 This is yet another example of Butler's 'tricky' essentialism; however, the notion that human males are wanderers is undermined by the character of Akin, who, against expectations, chooses to remain close to home.

28 'Construct' is the designation used to distinguish Jodahs – and later his sibling Aaor – as ooloi that combine both human and Oankali genetics. It is a rather problematic term, since it implies that the children from the human-Oankali genetic partnership are constructed – that is, produced in a way that is cultural rather than natural – while the children that the ooloi have among themselves prior to mating with humans are somehow produced in a 'natural' way. However, the information about Oankali reproduction provided in the novels suggests that all offspring are produced in the same way – through an ooloi selection of the genetic mix. The use of this term 'construct' suggests that Butler may have some residual sense of the body as purely natural *unless interfered with by technology* because, presumably, transspecies reproduction is not possible without technological intervention. Once the Oankali have evolved into whatever blend will emerge from human-Oankali mixing, the next generations begin to reproduce alone. Such reproduction is somehow considered more natural because it more closely resembles our sexual reproduction. Butler's use of this term in a trilogy in which she challenges the idea of the natural body suggests how easily unacknowledged assumptions about the body can emerge in discourse.

29 The 'it' refers to the neuter ooloi gender. However, it is tempting to think of Jodahs as a 'he' because it most commonly assumes a male form in response to the desire of a female mate. As the ooloi is the dominant partner in Oankali sexual couplings, the tendency to portray the neuter Jodahs as a male is perhaps further evidence of an underlying conservatism in the novels.

30 One of the benefits that the Oankali offer to the humans who live with them is correction for any diseases. For example, Lilith awakens in *Dawn* to discover the scar from an operation to remove a cancerous growth, and metastasis is prevented throughout Lilith's life by Nikanj's observation of her cells. In *Imago* the Oankali discover a community of humans who are able to reproduce themselves because a single female among them was fertile and became pregnant after a rape. This reproduction, inevitably, is completely in-bred, since all subsequent births within the community are from this woman and her offspring. A recessive genetic flaw in this stock – creating a disease of excessive bone growth – dooms these people to painful and disfigured lives without Oankali intervention.

## 3. Iain M. Banks: The Culture-al Body

1 Ultimately, the Mind in charge decides to leave earth alone, because it wants a control to ensure that the Culture has been doing the right thing when it chooses to interfere with other destructive civilizations.

2 This essay, published in 1994, will be referred to as 'Notes.' It is available from Rutger's SF Lover's archive at http://sflovers.rutgers.edu.

3 In a recent interview with *Science Fiction Weekly* no 274 (22 July 2002), Banks was asked whether the Culture was 'practically useful' as a guide to constructing a utopia, or 'simply an impossible goal, a make-believe Wonderland, utterly beyond our capacity.' Banks replied, 'practically useful for those who believe in reason in the same way that the idea of heaven is useful for those who have faith' (http://www.scifi.com/sfw/issue274/interview.html). The comparison to faith seems particularly telling here; Banks does not say that the ideals of a liberal humanist utopia are 'true' but rather that they are a useful blueprint for those who do believe in reason, the infinite perfectibility of mankind, etc., without offering any comment on his own belief or lack thereof.

4 Although I read the Culture novels as revealing a suspicion toward utopia, I think this critique exists in tension with Banks's desire to see the Culture as a utopia. In the *Science Fiction Weekly* interview, Banks says that his original idea for the Culture dates back to the first draft of *Use of Weapons*, and that he conceived of the Culture as 'unarguably [the] good guys,' allowing the narrative to focus solely on the morality of the non-Culture protagonist, Zakalwe (http://www.scifi.com/sfw/issue274/interview.html).

5 In the same *Science Fiction Weekly* interview, Banks suggests that he sees the critique of utopian impulses in the Culture novels more in terms of the requirements of good narrative than the inherent difficulties in establishing a utopia. When asked whether the Culture has become darker over the course of the seven novels, Banks replied, 'I always knew that what I was going to write about it would never in any sense faithfully represent what I was imagining, just because what I was imagining was a functioning utopia, and therefore pretty boring to read about. The interesting stuff was all at the far – and very thin – fringes, where the action and excitement was and dirty tricks could happen' (http://www.scifi.com/sfw/issue274/interview.html).

6 See Foucault, *The Order of Things* xviii.

7 On the surface, the Culture's decision to go to war to end Idiran imperialism seems a clear-cut case of justified intervention. However, the historical appendix to the book requires us to consider that the Culture's motives are mixed at best, since it is impossible to separate self-interest from altruism.

Represented as an extract from a book entitled *A Short History of the Idiran War*, the appendix points out that the Culture's 'sole justification for the rather relatively unworried, hedonistic life its population enjoyed was its good works' (484) and that when faced with the Idirans, the Culture had to either admit that it was more interested in its citizens' own comfort than in moral certitudes or wage war against the Idirans at any cost.

8 It does seem somewhat ironic that Horza would be so adamantly opposed to the Culture's body modifications, since his own body is continually moving between one form and another as he changes himself to impersonate various others. Horza believes that his organic ability is natural and contrasts it to the Culture's body enhancements. However, a Culture character has an insight into Horza's motivations that demolishes Horza's construction of the contrast between the natural body and the cultural body. Speaking to Horza in her mind, this character ponders, '*Who are you? What are you? A weapon. A thing made to deceive and kill, by the long dead. The whole subspecies that is the Changers is the remnant of some ancient war, a war so long gone that no one willing to tell recalls who fought it, or when, or over what. Nobody even knows whether the Changers were on the winning side or not. But in any event, you were fashioned, Horza. You did not evolve in a way you would call "natural"; you are the product of careful thought and genetic tinkering and military planning and deliberate design . . . and war; your very creation depended on it, you are the child of it, you are its legacy. Changer change yourself . . . but you cannot, you will not. All you can do is try not to think about it. And yet the knowledge is there, the information implanted, somewhere deep inside. You could – you should – live easy with it, all the same, but I don't think you do . . . And I'm sorry for you, because I think I know now who you really hate*' (*Consider Phlebas* 362–3, italics and ellipses in original).

9 The Azad style of reproduction has interesting parallels to the relationship to reproduction that can be produced through assistive reproductive technologies. In Azad, 'The dominant species is humanoid, but, very unusually – and certain analyses claim that this too has been a factor in the survival of the empire as a social system – it is composed of three sexes ... The one on the left ... is a male, carrying the testes and penis. The middle one is equipped with a kind of reversible vagina, and ovaries. The vagina turns inside-out to implant the fertilised egg in the third sex, on the right, which has a womb. The one in the middle is the dominant sex' (*The Player of Games* 74). This relationship, in which the genetic mother of the offspring is the apex – whose ovaries produce the egg – while the gestational mother of the offspring is the female – in whose womb the fetus grows – can be mapped to surrogate parenting contracts in which the surrogate mother provides only gestational 'service' for a fetus created from combining an egg and sperm

recovered from the contracting parents. In discussions around surrogate parenting and other ARTs, the contracting parents express concern regarding the 'quality' of the surrogate mother if her egg is to be used in reproduction, but are typically less concerned with the 'quality' of the surrogate mother if she will not be contributing any genetic material to the child (see Farquhar 35–41). This belief that the genetic contribution to the child is more significant than the social environment in which the child gestates is another example of genetic essentialism. Banks describes Azad culture as having believed, until recently, that the female sex was irrelevant to reproduction because they contributed only the gestational container, and he links this belief to other examples of their irrational gender prejudice: 'You know for millennia females were thought to have no effect on the heredity of the children they bore? They've known for five hundred years that they do; a viral DNA analogue which alters the genes a woman's impregnated with. Nevertheless, under the law females are simply possessions. The penalty for murdering a woman is a year's hard labour, for an apex. A female who kills an apex is tortured to death over a period of days' (*The Player of Games* 204).

10  Some examples: '"Now Level Three," the drone said … The screen held his gaze … The screams echoed through the lounge, over its formseats and couches and low tables; the screams of apices, men, women, children. Sometimes they were silenced quickly, but usually not. Each instrument, and each part of the tortured people, made its own noise; blood, knives, bones, laser, flesh, ripsaws, chemicals, leeches, fleshworms, vibraguns, even phalluses, fingers and claws; each made or produced their own distinctive sounds, counterpoints to the theme of screams … "That one is live … it is taking place now. It is still happening, deep in some cellar under a prison or a police barracks." … "This is no special night, Gurgeh, no festival of sado-erotica. These things go out every evening"' (*The Player of Games* 209–10, fourth and fifth ellipses in original); and '"We gain a great deal of pleasure from knowing at what cost this music is bought. … each of those steel strings has strangled a man. … [the pipe is] a female's femur, removed without anaesthetic … The drums are made from human skin. …. you see, Gurgeh, one can be on either side in the Empire. One can be the player, or one can be … played upon' (*The Player of Games* 222, ellipses in original).

11  It is interesting to note that the text links Gurgeh's desire to dominate to his 'abnormal' sexuality – in the Culture's terms. He refuses to have sex with partners of the same gender, and has remained a male throughout his life. In refusing him, one of the women he propositions explains: '"I feel you want to … take me," Yay said, "like a piece, like an area. To be had; to be … possessed." Suddenly she looked very puzzled. "There's something very … I

don't know; primitive, perhaps, about you, Gurgeh. You've never changed sex, have you?" He shook his head. "Or slept with a man?" Another shake. "I thought so," Yay said. "You're strange, Gurgeh"' (*The Player of Games* 24, ellipses in original). Additionally, the majority of his lovers choose to become male shortly after being with him. This sexual predilection explains another part of Gurgeh's attraction to Azad, since it is a social system of great sexual discrimination.

12 See de Lauretis, *The Practice of Love* 299–309.

13 Hardesty notes that Zakalwe's ruthlessness, which allows him to make anything – even the bones of his beloved – into a weapon, has its analogue in the novel in the use that the Culture makes of the emotionally damaged Zakalwe. He becomes their weapon in a civil war started by the Culture. That they are using Zakalwe is made clear by the conclusion of the novel in which it is revealed that although he understood his mission to be winning the war, the Culture's true objective was simply that he bolster the strength of the side he was assigned to sufficiently so that the war might not end too quickly. Banks, however, argues in his interview with *Science Fiction Weekly* no. 274 (22 July 2002) that the Culture remains morally just despite its relationship with Zakalwe and people like him: 'Zakalwe is a weapon and the Culture uses him, but the deceit is all on Zakalwe's side (not that what he's kept secret would necessarily have excluded him from working for the Culture, though his fascination factor as a kind of moral grotesque would have been off the scale). He has responsibility for his own actions, no matter what. Special Circumstances needs people like Zakalwe for a very limited number of operations and it isn't allowed to just make them so it has to find them elsewhere, after the damage to them has been done (and done by somebody else)' (http://www.scifi.com/sfw/issue274/interview.html).

### 4. Cyberpunk: Return of the Repressed Body

1 Mark Dery's study of cyber-culture, *Escape Velocity*, suggests that if cyberpunk survives, it does so only in cultural practices, not in science fiction writing. While Dery's work provides some readings of cyberpunk texts in relation to the cultural practices of musicians and performance artists who characterize themselves as cyberpunk, all of the literary texts discussed are from the 1980s.

2 Suvin asks, if cyberpunk creates a 'structure of feeling' in Raymond Williams's terms, then for whom does it do so? He suggests that the answer is affluent, first-world youth and technicians/artists of the new communication media and argues, 'The dilemma of how personal actions and con-

duct relate to social change is simultaneously inescapable and insoluble within Gibson's model ... In sum, a viable this-worldly, collective and public, utopianism simply is not within the horizon of the cyberpunk structure of feeling' (358).

3  See Ross, 'Cyberpunk in Boystown.' See also Nicola Nixon, who in 'Cyberpunk: Preparing the Ground for Revolution or Keeping the Boys Satisfied?' succinctly states the various complaints that feminist critiques have brought to the debate. Nixon describes the sub-genre as one that lacks a political agenda, for all its claims to be subversive, and she points to misogyny as the ground for many of its tropes. Nixon links the repression/rejection of the body in cyberpunk to the misogynistic tradition of Cartesian dualism, which associates the body with reviled femininity and posits mastery and transcendence as the qualities of the masculine mind. As Nixon convincingly demonstrates, the imagery of 'penetrating' defences and 'riding' programs used to describe cyberspace runs suggests that cyberpunk heroes are successful because of 'their facility, in short, as metaphorical rapists' (234). Nixon further argues that the history the sub-genre creates for itself, a history which admits to influences from male New Wave and hard SF authors but omits the influence of feminist SF from the 1970s, is further evidence of its repression of what has been feminized. Samuel Delany also comments on this denial of the 'mothers' of SF in 'Some *Real* Mothers.'

4  The analysis of MUD (multiple user domain or dungeon) communities by Lori Kendall also supports the assessment that the opportunity to play with gender roles does not lead to an understanding that gender is always a performance. Instead, as Kendall points out, the fact that people can gender cross-dress in a MUD tends to create increased emphasis on discovering the true gender that underlies the performance. Kendall writes: 'The stereotypes of masculine and feminine identity found on MUDs aren't new. Nor is the higher value placed on the "masculine" characteristics of intelligence and aggressiveness. But the greater male presence online and the limitations of this form of textual communication create a context in which these stereotypes are relied upon to a greater degree. So the answer to the question with which I began my research is that gender, in fact, has a great deal of meaning online. Although individuals can choose their gender representation, that does not seem to be creating a context in which gender is more fluid. Rather, gender identities themselves become even more rigidly understood. The ability to change one's gender identity online does not necessarily result in an understanding that gender identity is always a mask, always something merely performed. Rather, there can be an increased focus on the "true" identities behind the masks. Further, what I've found is that the standard

expectations of masculinity and femininity are still being attached to these identities' ('MUDder,' 221–2).

5  Rosi Braidotti, 'Cyberfeminism with a Difference.' Available at www.let.ruu.nl/women_studies/rosi/cyberfem.htm.

6  See Turkle 219–21. Turkle does use aliases for her interview subjects to hide their identities, but she does not give any indication whether or not her choice of the name Case is related to her reading of *Neuromancer.*

7  Mark Dery cites a radio interview with Gibson that supports my reading of this text. In the interview, Gibson says that the novel is a working through of 'some ideas I'd gotten from reading DH Lawrence about the dichotomy of mind and body in Judaeo-Christian culture' (interview with Terry Gross on 'Fresh Air,' National Public Radio, 31 August 1993, cited in Dery 248). Thomas Foster also argues that 'cyberpunk does not simply devalue the body but instead also foregrounds and interrogates the value and consequences of inhabiting bodies' ('Meat Puppets' 11). While I agree with Foster's assessment that the consequences of inhabiting bodies are addressed by cyberpunk, I disagree with the reading of Gibson's text that he goes on to make in his article. Foster argues that the novel is about the parallels that can be seen between humans and machines in their ability to overcome the limitations of their 'programming' and restructure themselves. In the novel Wintermute overcomes the limits of how it is 'wired', which Foster links to Molly's continual reference to the way she is 'wired,' to explain her choices. Foster argues that since Wintermute is able to overcome the constraints of its programming, so, too, are humans able to overcome their own constraints. I agree with this sentiment, but do not agree that Foster's reading of the novel demonstrates that Case and Molly have overcome their programming, as he suggests. He argues that Case overcomes the limitations of his self-destructive behaviour, and Molly overcomes the limitations of her feminine cultural role as meat puppet. However, Case is self-destructive *because* he can't get into cyberspace, and the tendency partially goes away when he gets his elite life back, not because he has been able to reprogram himself into valuing material existence. Further, he still takes drugs the night before the big run, suggesting that his destructive behaviour remains intact. Molly continues to use her body and sell her skills to the project of her employer, though in the role of assassin/bodyguard rather than as meat puppet. However, she remains detached from other human beings, treating them as objects rather than as subjects, so I would argue that her basic 'programming' has not changed *that* much. In fact, she retreats from a continuing relationship with Case after the 'job' is over precisely with the excuse that it is just the way she is wired.

8  The Sprawl is the 'wrong side of the tracks' in an urban centre that runs

from Boston to Atlanta, also referred to as BAMA, the Boston Atlanta Metropolitan Axis.

9  N. Katherine Hayles argues that the remainder of the trilogy does not sustain this privileging of material over virtual reality. Instead, she suggests, 'As the arc of the trilogy progresses, the preponderance of evidence shifts to support the claims [made by Winternute that there is no difference between 'real' life and cyberspace life], however much Case resists it initially' ('How Cyberspace Signifies' 115).

10  And although Case is shown to refuse the virtual Linda Lee and the life he could have with her, his escape into cyberspace is linked to his desire to avoid a reality in which he has lost her to death: 'Once he woke from a confused dream of Linda Lee, unable to recall who she was or what she'd ever meant to him. When he did remember, he jacked in and worked for nine straight hours' (*Neuromancer* 59).

11  See Larry McCaffery, 'An Interview with William Gibson' 280.

12  For examples see Heim (especially *Metaphysics*) and Dewdney.

13  Mark Slouka argues most forcefully against this tendency to cut off connections to real people in our physical vicinity in *War of the Worlds*. While I agree with Slouka's argument that cyberspace can tend to distract us from material problems and encourage us to turn our attention elsewhere, I disagree with the larger thrust of his work. Slouka's general argument is that the blurring of boundaries between real life (RL) and virtual reality has the consequence of undermining the concept of truth. He ultimately argues that we can return to an ethical world 'by resuscitating our faith in truth in general, by recognizing its importance, by rededicating ourselves to its pursuit' (149) and that respecting the distinction between real reality and virtual reality is the beginning of this work. I, too, argue that a focus on material, embodied reality is necessary for ethics, but I want to argue for a focus on the specificity of material reality, an attention to its detail that precludes the erasures and oppressions that its abstraction to data can produce. Slouka's reliance on an unproblematized notion of truth in fact reproduces some of the problems of abstraction that I associate with a cyberspace world. In refusing to recognize that the social world is a construct that may be seen differently from different perspectives, Slouka inadvertently aligns himself with the totalizing logic of certain cyberspace enthusiasts.

14  One of the problems with the contrast between Gina and Mark is that the assumptions underlying their different attitudes leave intact the gender and Cartesian binary that associated woman with body and man with abstract mind. Further, as Jenny Wolmark argues, traditional gender roles are also reinforced by representing the male as cyberpunk hero and the female as

mothering/nurturing figure (125). Although I would agree with Wolmark to the extent that Mark is clearly the celebrated partner in their musical collaboration and Gina clearly devotes more of her energy to nurturing Mark and his career than to accomplishing her own goals, I do not see Mark as the cyberpunk hero in the novel. Instead, Sam (short for Cassandra) is the main focus of the hacker sub-plot and the one who ultimately saves the Net, reversing the gender expectations about cyber-cowboys.

15 In 'The Psychodynamic Effects of Virtual Reality' Leslie Harris explains that the experience of being disoriented in 'real reality' is a cognitive side-effect of long sessions in simulated environments. He describes the role that past experience has in forming our perceptions; we compare current perceptions to past experience and – if possible – fill in the details and expectations from experience. Past experience makes up a series of possible worlds against which we view the actual world, and our past experiences in virtual worlds are treated by the body no differently from our past experiences in 'real' life. Harris describes how his own experience of time spent in simulated driving games later creates a sense of disorientation when he is confronted with the controls of a physical car. His work draws on the role of the body in learning and the way in which proficiency in a task is achieved by responses becoming intuitive – hard-coded as it were – based on past experience. In 'The Challenge of Merleau-Ponty's Phenomenology of Embodiment for Cognitive Science,' Hubert L. Dreyfus and Stuart E. Dreyfus explain that the embodied nature of learning has created obstacles for the development of artificial intelligence. Drawing on the same insight regarding the body's past experiences structuring current perceptions of the world, they describe bodily constraints on how we generalize learning. The architecture of the brain limits paths of inputs/outputs, creating a body-dependent order of presentation of external stimuli. Researchers have discovered that although neural nets can learn skills through the same process as brains do, they do not learn to respond in a 'human' way to the same experiences because they do not experience stimuli in the same body-dependent order of presentation. Taken together, these two articles suggest that the phenomenological experience of exceeding the body's 'limitations' that is possible in virtual space may work to create radically altered human subjectivities.

16 The name is similar to the virtual character Arthur Fishell who was created by Atari Labs staff in the early 1980s. Allucquère Stone describes the ways in which the lab staff used the character of Fishell to work on their ideas for interactivity between computers and humans. They created a virtual presence for this 'person' primarily using email. At one point in time, he was

even named pro tem director of the lab. Atari staff outside the lab group came to believe that Arthur was a 'real' person (see *The War of Desire* 139–43).

17  See 'Unstable Networks' for Sterling's reassessment of the subversive potential of cyberpunk. He has retreated from his earlier enthusiasms about the 'the subversive potential of the home printer and the photocopier' (Preface to *Mirrorshades* xiv). Sterling now realizes that technology alone cannot solve social problems, observing, 'Is our technology really a panacea for our bad politics? I don't see how. We can't wave a floppy disk like a bag of garlic and expect every vampire in history to vanish' ('Unstable Networks' 36).

18  Laura Cherniak argues in 'Pat Cadigan's *Synners*: Refiguring Nature, Science and Technology' that the structure of the narrative itself reinforces the importance of communal cooperation to the resolution of the plot: 'the enigma is not solved by the actions and thoughts of a single person, piecing it together like a detective. Instead, the characters do not even know each other. However, collectively, knowingly or not, they move the narrative forward and solve the enigma' (123). Cherniak's argument, like mine, reads Cadigan's concern with technology as predominantly an ethical engagement.

19  Cadigan has stated in an interview at the 1996 Virtual Futures conference that her concern with technology in her novels is always a concern with the interaction between technology and human beings, and with changed interaction among human beings as a consequence of technology: 'Mostly, what I'm concerned with, if you were to distill it down into what they call "high-concept" in the States, is the impact of technology on human beings. What kind of culture grows up around it, what kind of beliefs grow up around it, even what kind of superstitions and rituals grow up around it? What sort of unforeseen effects does it have?' See Miss M., 'An Interview with Pat Cadigan.'

20  Cherniak points out that Cadigan's version of cyberpunk shows a complex interaction between technoscience and capitalism, not simply technoscience as the tool of capitalism as Jameson suggests it is in 'Cognitive Mapping.'

21  Cherniak sees in the shift from 'sinners' to 'synners' a rejection of the pastoral ideal that Gabe tries to enact by isolating himself from technology. Instead, she argues that all that is possible are the kind of cyborg subjectivities theorized by Haraway (Cherniak 82). Although I would agree to the extent that Cadigan clearly suggests that it is impossible to go back to a time before the technology, I think that she also emphasizes that it is important to decide how to use – or not use – certain technologies. Thus, I don't see Gabe's isolation as an attempt to return to a pastoral, but instead as a con-

scious decision to value the friendships he can establish in the material world over those he makes in virtual reality. The way his relationship with Gina is presented as a 'healthy' alternative to his cyber-relationships with Marly and Caritha reinforces this reading. When Gabe is connected to technology, he is connected to only a solipsistic virtual simulation in which parts of his own self become other characters. He is far more isolated (and hence less cyborg – since the cyborg is about connections, fusions, and the crossing of boundaries) when he is connected to technology than when he rejects it. It is important to remember that in the cyborg myth Haraway argues that the metaphor of the cyborg is necessarily impure and connected to multiple types of subjectivities as a way to understand our multiple connections with *human beings* in the world who are not identical to us (that is, for a politics based on affinity and connection instead of 'pure' identity of race, gender, etc.).

22  In the interview given at the 1996 Virtual Futures conference, Cadigan said, 'I believe that any tool is only as good as the people that use it ... What I wanted to show in *Synners* was that in some cases cyberspace or whatever you choose to call this technology, will bring people together and in other cases it will divide people, it will isolate people. Again, it always depends on how it is used. It is only as good as the person or people using it' (Miss M., 'An Interview with Pat Cadigan'). This statement emphasizes not only the importance of community and connection among people to Cadigan, but also her refusal to engage in a polemical critique of technology which insists on reading a technology as 'good' or 'bad' without any reference to the context within which it is used.

23  See Lowe 33–5.

24  Haraway herself has commented on the failure of the cyborg as a cultural figure to represent the myth she argues for in the 'Cyborg Manifesto.' She discusses some of her rethinking of the cyborg and rearticulates the politics she was describing in the original essay in 'Cyborgs at Large,' an interview with Constance Penley and Andrew Ross, and in 'The Actors Are Cyborg, Nature Is a Coyote, and the Geography Is Elsewhere,' both found in the *Technoculture* collection edited by Penley and Ross.

25  See Miss M., 'An Interview with Pat Cadigan.'

## 5. Raphael Carter: The Fall into Meat

1  Mirabara is Keishi's last name. Maya's insistence on using this more impersonal form of address suggests the distance she wants to maintain between the current incarnation and the 'real' person.

2 This is the pronoun that Carter uses to describe zirself. Zir's discussion of gender identity can be found on zir's homepage at www.chaparraltree.com. Carter prefers to use the term 'epicene,' which zie defines as 'partaking of the characteristics of both sexes,' to describe zir gender identity.

3 Keishi offers a very good refutation of Maya's position, but one that, ultimately, is not endorsed by the novel in its support of Maya's decision. The novel is written as Maya's reflection back on events of the time and she never suggests that she regrets her choice. Keishi argues: 'If you take flesh as your starting point ... you're always going to find some way that silicon falls short. But there's nothing special about flesh. Look, sex wasn't invented by some loving God who wants us all to understand each other and be happy. It was made by nature, and nature doesn't give a damn whether our hearts hook up or not, just as long as our gametes do. Why should evolution get to make all the decisions? Why can't we use something that *is* designed to bring people together? If you turn the comparison around, and start with cabling, then love in the meat starts to look pretty shabby. Love happens in the mind, in the soul – what does the union of two sweating bodies have to do with that?' (158–9). Keishi's argument relies on mind/body dualism ('love happens in the mind'), a fragmentation of self that the novel rejects in its insistence on embodiment as an essential aspect of being human.

4 The novel even points to the role of language in shaping our value systems and constructing hierarchies. Keishi tells Maya that the tool she is given by the free African technology Net to fight the repression of homosexuality on the Net in the Fusion (where they live) is the ability to speak Sapir – a computer language – as her first language. The social programming accomplished by the Russian language is removed from her subjectivity: 'human language protects the mind from Sapir. That's why the Africans are so far ahead – don't you see? They don't get their Sapir from a fluency chip. They get their first brainmod at one year, and learn it as a first language. It's the lingua franca of their continent, and that makes all the difference. The first computer languages were pidgins, formed the same way any pidgin is formed, by a dominant race thrusting its words onto the grammar of a subordinate one. Then with KRIOL, the pidgin acquired native speakers. In Sapir the tables are turned – it changes human thought to fit computers, not the other way around' (270). This quotation suggests that – unlike Cadigan – Carter believes that becoming more like machines, thinking as the computer does, is socially beneficial. I suspect that this view is linked to the idea that prejudice is irrational and machines are grounded in rationality. However, as I will argue in more detail below, I believe that reducing human subjectivity to rationality alone is socially and psychologically damaging. It would seem

that Carter, too, shares this perspective to some degree, given the representation of Maya's rejection of Keishi because the relationship would be one of minds only. There is also an interesting parallel between the notion of Sapir as something that can change the world beyond computing and the language Forth, which was very popular in the late 1970s. Allucquère Stone describes the attitudes surrounding Forth: 'Forth wasn't simply a programming language. Rather, it was an element within a system, a building block in an associated group of elements that, taken together, represented a complete philosophy of life. This way of life, if practiced diligently, would effect a gradual but pervasive transformation not only in the life of the practitioner, but in the greater world as well' (*War of Desire* 105). It is also important to note that while Carter's vision suggests that we might change for the machines in learning a computer language as our first language and thus have our perceptions of the world shaped by a logic other than the one embedded in human languages, zie does not endorse the idea that our minds can simply become programs that might be uploaded into various media without change. Human language, rather than human embodiment, is the source of bias that must be transcended in Carter's vision.

5  See Ussher, 'Framing the Sexual "Other"' 140.

6  The Fusion is not all that clearly defined but seems to be a European-Asian economic and political merger. It is always contrasted with the other major global power, Africa.

7  Mark Slouka's *War of the Worlds* shares some of this tendency, although his analysis is largely focused on the consequences of existing technology. However, the distinction between science fiction and social critique is blurred in his work by passages such as the following: 'Our home computers, to take just one example, will soon come with a face capable of responding to our expressions, understanding our gestures, even reading our lips. Its eyes will follow us around the room. We'll be able to talk with it, argue with it, flirt with it. We'll be able to program it to look like our husband or our child. Or the Holy See, I suppose. Will it have emotions? You bet. Scream at it and it will cower or cringe' (8).

8  See Richard Kadrey, 'Reach Out and Touch Someone.' Around the same time, the *National Post* also published a column about cyber-infidelity and the problems (ethical and pragmatic) of determining if one's spouse is cheating or not by participating in cybersex encounters (see Patricia Pearson, 'Is That a Mouse in Your Pants …').

9  Available at www.teledildonics.com.

10  See www.ifriends.com

11  It is true that in subsequent novels in the series we are told that Case got mar-

ried, suggesting that change is possible. However, Case is also not present in these novels other than via this reference, which suggests that as a connected subject he no longer has anything to offer cyberpunk narratives.

## 6. Jack Womack and Neal Stephenson: The World and the Text and the World in the Text

1  It is interesting to note that both of these texts about adolescent girls are written by male authors. It is beyond the scope of the current project to engage with this issue. Octavia Butler's *Parable of the Sower* is in the form of a young girl's diary, and so offers a comparative text written by a woman. Both Michael Levy and Joan Gordon have explored the relationship between *Parable of the Sower* and *Random Acts of Senseless Violence* and the question of the diary form.

2  There is also a second 'edition' of the *Primer* produced in the novel that differs from the original *Primer* in that it does not use live ractors. I will discuss this second edition of the *Primer* in more detail below.

3  The ractor, Miranda, who does most of the speaking parts for the copy of the *Primer* used by Nell comes to feel as if she is raising Nell through the *Primer*. Although Miranda cannot speak lines other than those provided by the *Primer*'s script, she believes that she can influence the effect that the text will have on Nell through her delivery of them. Miranda becomes so attached to Nell and concerned with continuing her parenting that she sacrifices her own career as a ractor in order to take the *Primer* jobs over other jobs. Miranda changes her work schedule to ensure that she is available during the hours when Nell is most likely to use the *Primer*.

4  I will return to this idea of agency and resistance in relation to both novels at a later point in this chapter after an initial discussion of *Random Acts*.

5  This novel fits within a science fictional framework because it is part of a series of novels by Womack set in this imagined future. The other novels in the series are more typical of the SF genre.

6  The term 'queer' is used by Lola throughout the text in her efforts to understand her identity as a sexual being. Lola both reports 'queer' as the pejorative term used by her peers in their rejection of her, and uses the term as a self-descriptor in her attempts to articulate her feelings in her diary. The gap between representation and experience, which I will discuss further below, is crucial to Lola's attempts to work through what it means to be queer. She lacks another term to describe her desire, and so must struggle to reconcile her recognition that the term is used pejoratively by her peers with her own sense of love and desire. I will discuss Lola's queerness in more detail below.

7 Joan Gordon argues in 'Two SF Diaries at the Intersection of Subjunctive Hopes and Declarative Despair' that *Random Acts* branches out from the aborted attempt in Anne Frank's diary to believe that 'people are truly good at heart.' She reads both *Random Acts* and Butler's *Parable of the Sower* as exploring the space remaining to these young girls to survive as colonized subjects of fascistic future regimes. Gordon concludes that diaries are by nature a subjunctive form because they are open-ended, always leaving a space for hope and change. Gordon's reading of *Random Acts* emphasizes what is positive in Lola's relationship with her diary. In contrast, my reading focuses on the tension between discursive and non-discursive self-presentation and the difficulties Lola experiences because she cannot sustain hope or accomplish change outside the written text.

8 The 'ghetto' is how Lola's old friends, her private-school teachers, and her right-wing aunt refer to their new neighbourhood.

9 See Freeman 42–9.

10 Boob is the family nickname for Lola's nine-year-old sister, Cheryl. Lola's family nickname is Booz.

11 The choice to use 'verb' as a verb is an example of exactly this type of language use.

12 Some examples of slang that emphasizes action in the text are: 'it's handleable,' 'blade you,' 'lipstill' (all from 106); 'bedding with us,' 'churching,' 'fastfoot' (all from 118); and 'truth me' (133).

13 Given the age of the characters, 'boys and girls' seems more appropriate than 'men and women.' However, although the incidents are not described in the novel, there is a suggestion that Jude sleeps with men, pretending to be ten years older than she is. Jude trades sex acts for material goods, and it is her sexual connections that provide the limousine that rescues Jude and Iz from the riot.

14 A wonderful image of the fact that real power lies beyond the reach of revenge of the dispossessed is given in the description of individuals trying to attack the helicopters that are removing important people from the reach of the rioting crowd: 'Copters flew over and people threw rocks and bricks in the air but they didn't hit the copters, only the people who threw them' (177). The image of the limousine that rescues Jude and Iz indiscriminately driving over rioters who block its escape also fits within this pattern.

15 Note that Hackworth, who designed the *Primer*, is a software engineer and is frequently referred to simply as an engineer. Ultimately, he is referred to as The Alchemist, suggesting the power of discourse to change social status, to turn dross into gold through appropriate discursive positioning.

16 The matter compiler or MC is one of the main technologies in *The Diamond*

*Age*. The MC creates any desired object from its constituent atomic parts based on software plans written by engineers like Hackworth. The MC technology is related to nanotechnology, the other main technology in the novel.

17 The changed typeface is a convention used in Stephenson's text to demarcate those passages that are 'read' from the *Primer* from the rest of the text.

18 The fate of the Asian girls in *The Diamond Age* – called alternately the Mouse Army and the Disenchanted Army in the text – would also support this perspective. The characters who rescue the girls from death believe that they have failed to raise them properly because resources required that they be raised primarily by the books. Dr X, the person responsible for the plan to save them, initially states that it was a mistake to do so. He corrects himself by explaining, 'It would be more correct to say that, although it was virtuous to save them, it was mistaken to believe that they could be raised properly. We lacked the resources to raise them individually and so we raised them with books. But the only proper way to raise a child is within a family. The Master [Confucius] could have told us as much, had we listened to his words' (455).

19 Freeman comes to a similar conclusion about writing the self as work to change the world (see 223–31).

20 The Fist are a group of young men from the Celestial Kingdom who are motivated by racism to attack and subjugate the Leased Territories.

21 See Butler, *Excitable Speech*, especially 127–63, for a further elaboration of this idea. Butler's basic argument, drawing on Austin, is that some representations can function as illocutionary speech acts – that is, they can enact what they state simply in the statement – while other speech acts fail to be successful as illocution. The difference lies in whether or not the person performing the speech resides within a position of authority.

## Conclusion

1 Available at http://www.ugcs.caltech.edu/~phoenix/vinge/vinge-sing.html, accessed on 19 September 2001.

2 The definition and 'The Extropian Principles' (which I will discuss in more detail below) are written by Max More and are available at www.extropy.org/extprn3.htm.

3 I use the word 'technology' here in the sense that the Extropians themselves see these various engagements with body modifications as technologies, and also to invoke the use of the word to describe institutional practices and discourses that discipline the body to produce the subject in specific ways (as Foucault, Balsamo, and de Lauretis have used the term). I do not intend to suggest that each of these technologies is equally plausible or equally

accepted within the mainstream scientific community. For example, the technology of mind-uploading seems to be a science fiction trope and Extropian dream only, not a path of research currently being pursued.

4 This description is available at www.mit.edu/people/jpbonsen/jpbonsen-home.html.

5 The gender-specific language here is an intentional choice.

6 See Clynes and Kline 29–33.

7 Available at www.extropy.org/faq.

8 See www.extropy.org/faq/minority.html.

9 See www.extropy.org/faq/elite.html.

10 See 'Cyborgs at Large,' in Penley and Ross, *Technoculture* 16.

11 Available at www.extropy.org/eo/.

12 For examples of sites that include reading lists, see www.extropy.org/faq/topics.html which provides a list of 'required' reading that one is expected to master before asking 'obvious' questions on the mailing list; www.aleph.se/Trans/index.html, which includes a mixture of fictional and non-fictional texts under each of its various Technology links; and a reading list which distinguishes between fictional and non-fictional sources found following the Extropian Principles at www.extropy.org/extprn3.htm.

13 My thinking on the posthuman has been influenced by Hayles, but my concern is less with tracing how the history of cybernetics leads to a disembodied view of the subject and information (the focus of Hayles's book) and more with attempting to think through how we might develop another concept of the posthuman that is attentive to embodiment.

14 See Weiss, *Body Images* 162.

15 See Haraway, 'Situated Knowledges' 190.

16 See Veronica Hollinger, 'Women in SF and Other Hopeful Monsters,' for her discussion of this tendency of SF to push the limits of intelligible subject positions in terms of the ideological tensions between being a good 'woman' and being a good 'human subject.'

17 Steven Best and Douglas Kellner argue something similar when they claim that the posthuman 'signifies the end of certain misguided ways of conceiving of human identity and the nature of human relations to the social and natural environments, other species, and technology' (271) and they argue for what they call a '*postmodern humanism*' which would avoid such problems. There are two key differences between their position and mine. First, although they call for an end to anthropocentrism and the illusion of human sovereignty over nature, they don't focus particularly on the erasure of embodied specificity as one of the roots of these problems, nor do they offer a way to conceive of human nature in other terms. Second, my vision of ethi-

cal posthumanism sees some of the continued assumptions of the discourse of humanism itself as part of the problem and thus I would argue that a humanism informed by postmodern thought is not sufficient to overcome some of the very limitations Best and Kellner draw attention to, since it is the discourse of humanism itself that sets humans apart from the rest of the world and thus contributes to the very divides they name as the problem.

# Bibliography

Adam, Alison, and Eileen Green. 'Gender, Agency, Location and the New Information Society.' Loader, *Cyberspace Divide* 83–97.

Alaimo, Stacy. 'Displacing Darwin and Descartes: The Bodily Transgressions of Fielding Burke, Octavia Butler, and Linda Hogan.' *Isle Interdisciplinary Studies in Literature and the Environment* 3.1 (1996): 47–66.

Aldiss, Brian. *The Detached Retina: Aspects of Science Fiction and Fantasy.* Syracuse University Press, 1995.

Aldridge, Susan. *The Thread of Life: The Story of Genes and Genetic Engineering.* Cambridge University Press, 1996.

Althusser, Louis. 'Ideology and Ideological State Apparatuses.' *Contemporary Critical Theory.* Edited by Dan Latimer. New York: Harcourt Brace Jovanovich, 1989. 1–60 [originally from *Lenin and Philosophy and Other Essays*, 1969].

Appleyard, Bryan. *Brave New Worlds: Staying Human in the Genetic Future.* Viking, 1998.

Attebery, Brian. *Decoding Gender in Science Fiction.* Routledge, 2002.

Atzori, Paulo, and Kirk Woolford. 'Extended Body: An Interview with Stelarc.' Kroker and Kroker, *Digital Delirium* 194–9.

Aurigi, Allesandro, and Stephen Graham. 'The "Crisis" in the Urban Public Realm.' Loader, *Cyberspace Divide* 57–80.

B.K. 'Are You My Daddy?' *Elm Street Magazine*, September 1999, 26.

Balsamo, Anne. 'Feminism for the Incurably Informed.' *SAQ* 92.4 (Fall 1993): 681–712.

– 'Forms of Technological Embodiment: Reading the Body in Contemporary Culture.' *Cyberspace/Cyberbodies/Cyberpunk.* Edited by Mike Featherstone and Roger Burrows. Sage, 1995. 215–37.

– *Technologies of the Gendered Body.* Duke University Press, 1996.

Banks. Iain M. *Consider Phlebas.* Bantam Books, 1987.

- *Excession.* Orbit, 1997.
- Interview in *Science Fiction Weekly* no. 274 (22 July 2002). http://www.scifi.com/sfw/issue274/interview/html.
- *Look to Windward.* Orbit, 2000.
- *The Player of Games.* Orbit, 1988.
- 'Some Notes on the Culture.' *Rutgers's SF Lover's Archive.* 1994. http://sflovers.rutgers.edu. 19 April 1999.
- 'The State of the Art.' *The State of the Art.* Orbit, 1991.
- *Use of Weapons.* Bantam Books, 1990.
Barglow, Raymond. *The Crisis of Self in the Age of Information: Computers, Dolphins, Dreams.* Routledge, 1994.
Barnard, Jeff. 'Talk about Building a Better Mousetrap; This Gene Experiment Created Smarter Mice.' *Edmonton Journal,* 2 September 1999, A1.
Barr, Marleen. *Feminist Fabulations: Space/Postmodern Fiction.* University of Iowa Press, 1992.
Barrett, Michèle. 'Feminism and the Definition of Cultural Politics.' *Feminism, Culture and Politics.* Edited by Rosalind Brunt and Caroline Rowan. Lawrence and Wishart, 1982. 37–58.
- *The Politics of Truth: From Marx to Foucault.* Stanford University Press, 1991.
- 'Words and Things: Materialism and Method in Contemporary Feminist Analysis.' *Destabilizing Theory: Contemporary Feminist Debates.* Edited by Michèle Barrett and Anne Philips. Stanford University Press, 1992. 201–19.
Baudrillard, Jean. *Simulations.* Foreign Agents Series. Translated by Paul Foss, Paul Patton, and Phlip Beitchman. Semiotext(e), 1983.
- 'The Year 2000 Has Already Happened.' Translated by Nai-fei Ding and Kuan-Hsing. *Body Invaders: Panic Sex in America.* Edited by Arthur Kroker and Marilouise Kroker. CultureTexts. St Martin's Press, 1987. 35–44.
- 'Two Essays.' *Science Fiction Studies* 18.3 (1991): 309–20.
Beal, Frances. 'Black Women and Science Fiction Genre: Interview with Octavia Butler.' *Black Scholar* 17.2 (1986): 14–18.
Belkin, Lisa. 'It's a ~~Girl/Boy~~ or Whatever You Want.' *Edmonton Journal,* 15 August 1999, F1.
Best, Steven, and Douglas Kellner. *The Postmodern Adventure: Science, Technology and Cultural Studies at the Third Millennium.* Guilford Press, 2001.
Bordo, Susan. *Unbearable Weight.* University of California Press, 1993.
Boulter, Amanda. 'Polymorphous Futures: Octavia E. Butler's *Xenogenesis* Trilogy.' *American Bodies: Cultural Histories of the Physique.* Edited by Tim Armstrong. New York University Press, 1996. 170–85.

Bourdieu, Pierre. *Distinction: A Social Critique of the Judgement of Taste.* Translated by Richard Nice. Cambridge: Harvard University Press, 1984.

Braidotti, Rosi. 'Cyberfeminism with a Difference.' *Women's Studies at Utrecht Homepage* 3 July 1996. www.let.ruu.nl/women_studies/rosi/cyberfem.htm. 24 February 1999.

Braidotti, Rosi, and Nina Lykke, eds. *Between Monsters, Goddesses and Cyborgs: Feminist Confrontations with Science, Medicine and Cyberspace.* Zed Books, 1996.

Broderick, Damien. *Reading by Starlight: Postmodern Science Fiction.* New York: Routledge, 1995.

– *The Spike: How Our Lives Are Being Transformed by Rapidly Advancing Technologies.* TOR, 2001.

Brosnan, Mark. *Technophobia: The Psychological Impact of Information Technology.* Routledge, 1998.

Brown, Carolyn. 'Utopias and Heterotopias: The "Culture" of Iain M. Banks.' *Impossibility Fiction: Alternativity Extrapolation Speculation.* Edited by Derek Littlewood and Peter Stockwell. Rodolphi, 1996. 56–74.

Bukatman, Scott. 'Postcards from the Posthuman Solar System.' *Science Fiction Studies* 18.3 (1991): 343–57.

– *Terminal Identity.* Duke University Press, 1993.

Butler, Judith. *Bodies That Matter: On the Discursive Limits of 'Sex'.* Routledge, 1993.

– *Excitable Speech: A Politics of the Performative.* Routledge, 1997.

– *Gender Trouble.* Routledge, 1990.

– *The Psychic Life of Power.* Stanford University Press, 1997.

Butler, Judith, Slavoj Žižek, and Ernesto Laclau. *Contingency, Hegemony, Universality: Contemporary Dialogues on the Left.* Verso, 2000.

Butler, Octavia. *Xenogenesis.* Guild Books. nd [combined edition of *Dawn*, originally published 1987, *Adulthood Rites*, originally published 1988, and *Imago*, originally published 1989].

Cadigan, Pat. *Synners.* Bantam Books, 1991.

Carter, Raphael. *The Fortunate Fall.* TOR, 1996.

Cherniak, Laura. 'Pat Cadigan's *Synners*: Refiguring Nature, Science and Technology.' *Feminist Review* 56 (Summer 1997): 61–84.

Cherniavsky, Eve. '(En)gendering Cyberspace in *Neuromancer*: Postmodern Subjectivity and Virtual Motherhood.' *Genders* 18 (Winter 1993): 32–46.

Clynes, Manfred E., and Nathan S. Kline. 'Cyborgs and Space.' Gray, *The Cyborg Handbook* 29–33.

Collins, Jim. *Uncommon Cultures: Popular Culture and Post-Modernism.* Routledge, 1989.

Cowan, Ruth Schwartz. 'Genetic Technology and Reproductive Choice: An Ethics for Autonomy.' Kevles and Hood 244–63.

Csicsery-Ronay, Istvan, Jr. 'Cyberpunk and Neuromanticism.' *Storming the Reality Studio*. Duke University Press, 1992. 183–93.

Davies, Tony. *Humanism*. Routledge, 1997.

Delany, Samuel R. 'The Future of the Body – and Science Fiction and Technology.' *The New York Review of Science Fiction* 48 (August 1992): 1, 10–13.

– *The Jewel-Hinged Jaw: Notes on the Language of Science Fiction*. Dragon Press, 1977.

– 'The Life of/and Writing.' *New York Review of Science Fiction* 26 (October 1990):1, 8–13.

– 'Is Cyberpunk a Good Thing or a Bad Thing?' *Mississippi Review* 47/48 (1988): 28–35.

– 'Some *Real* Mothers: An Interview with Samuel R. Delany by Takayuki Tatsumi.' *Science Fiction Eye* 1.3 (1988): 5–11.

– *Starboard Wine: More Notes on the Language of Science Fiction*. Dragon Press, 1984.

de Lauretis, Teresa. *Alice Doesn't*. Indiana University Press, 1984.

– *The Practice of Love: Lesbian Sexuality and Perverse Desire*. Indiana University Press, 1994.

– 'Signs of W[a/o]nder.' De Lauretis et al. 159–74.

– *Technologies of Gender*. Indiana University Press, 1987.

de Lauretis, Teresa, Andreas Huyssen, and Kathleen Woodward, eds. *The Technological Imagination: Theories and Fictions*. Coda Press, 1980.

Dery, Mark. *Escape Velocity: Cyberculture at the End of the Century*. Grove Press, 1996.

Dewdney, Christopher. *Last Flesh: Life in the Transhuman Era*. HarperCollins, 1998.

Dreyfus, Hubert L., and Stuart E. Dreyfus. 'The Challenge of Merleau-Ponty's Phenomenology of Embodiment for Cognitive Science.' Weiss and Haber 103–20.

Egan, Greg. 'Cocoon.' *Luminous*. Millennium, 1998. 107–40.

– 'Learning to Be Me.' *Axiomatic*. Millennium, 1995. 201–20.

– 'Mister Volition.' *Luminous*. Millennium, 1998. 89–106.

– 'Reasons to Be Cheerful.' *Luminous*. Millennium, 1998. 191–227.

Evenson, Brad. 'Personality Can Be Transplanted, Researchers Find.' *National Post*, 19 August 1999, A1.

Fanon, Franz. 'The Fact of Blackness.' *Anatomy of Racism*. Edited by David Theo Goldberg. University of Minnesota Press, 1990. 108–26.

Farquhar, Dion. *The Other Machine: Discourse and Reproductive Technologies*. Routledge, 1996.

Federici, Silvia. *Caliban and the Witch: Women, the Body and Primitive Accumulation*. Autonomedia, 2004.

Fiske, John. *Understanding Popular Culture*. Unwin Hyman, 1989.

Flax, Jane. *Thinking Fragments: Psychoanalysis, Feminism and Postmodernism in the Contemporary West.* University of California Press, 1990.

Foster, Frances. 'Octavia Butler's Black Female Future Fiction.' *Extrapolation* 23.1 (1982): 37–49.

Foster, Thomas. 'Incurably Informed: The Pleasures and Dangers of Cyberpunk.' *Genders* 18 (Winter 1993): 1–10.

– 'Meat Puppets or Robopaths?: Cyberpunk and the Question of Embodiment.' *Genders* 18 (Winter 1993): 11–31.

Foucault, Michel. *Discipline and Punish: The Birth of the Prison.* Translated by Alan Sheridan. Vintage Books, 1977.

– *The History of Sexuality.* Volume 1. Pantheon Books, 1978.

– *The Order of Things: An Archaeology of the Human Sciences.* Vintage Books, 1994.

– *Power/Knowledge.* Edited by Colin Gordon. Pantheon, 1980.

Freedman, Carl. 'Science Fiction and Critical Theory.' *Science Fiction Studies* 14.2 (1987): 180–200.

– *Science Fiction and Critical Theory.* Weseylan University Press, 2000.

Freeman, Mark. *Rewriting the Self: History, Memory, Narrative.* Routledge, 1993.

Freud, Sigmund. *The Ego and the Id.* Translated by James Strachey. W.W. Norton, 1961.

– *The Interpretation of Dreams.* Translated by James Strachey. Basic Books, 1955.

– *On Sexuality.* Penguin Books, 1977.

Fuss, Diana. *Essentially Speaking: Feminism, Nature and Difference.* Routledge, 1989.

Gallop, Jane. *Reading Lacan.* Cornell University Press, 1985.

– *Thinking through the Body.* Columbia University Press, 1988.

Garber, Marjorie. 'Spare Parts: The Surgical Construction of Gender.' *The Gay and Lesbian Studies Reader.* Edited by Henry Abelove, Michele Aina Barale, and David M. Helperin. Routledge, 1993. 321–36.

Gevers, Nick. 'Cultured Futurist Iain M. Banks Creates an Ornate Utopia.' *Science Fiction Weekly* 8.30 no. 274 (22 July 2002). http://www.scifi.com/sfw/issue274/interview.html. 23 July 2002.

Gibson, William. *Neuromancer.* ACE Books, 1984.

Gilbert, Walter. 'A Vision of the Grail.' Kevles and Hood 83–97.

Gordon, Joan. 'Two SF Diaries at the Intersection of Subjunctive Hope and Declarative Despair.' *Foundation* 72 (Spring 1998): 42–8.

Graham, Elaine. *Representations of the Post/Human: Monsters, Aliens and Others in Popular Culture.* Rutgers University Press, 2002.

Gray, Chris Hables. *Cyborg Citizen: Politics in the Posthuman Age.* Routledge, 2001.

Gray, Chris Hables, ed. *The Cyborg Handbook.* With the assistance of Heidi J. Figueroa-Sarriera and Steven Mentor. Routledge, 1995.

Greely, Henry T. 'Health Insurance, Employment Discrimination and the Genetics Revolution.' Kevles and Hood 264–80.

Green, Michelle Erica. '"There Goes the Neighborhood": Octavia Butler's Demand for Diversity in Utopias.' *Utopias and Science Fiction by Women: Worlds of Difference.* Edited by Jane Donawerth and Carol Kilmertin. Syracuse University Press, 1994. 166–89.

Grossberg, Lawrence, Cary Nelson, and Paula Treichler, eds. *Cultural Studies.* Routlege, 1993.

Grosz, Elizabeth. *Jacques Lacan: A Feminist Introduction.* Routledge, 1990.

– 'Refiguring Lesbian Desire.' *Race, Class, Gender, and Sexuality: The Big Questions.* Edited by Naomi Zack, Laurie Shrage, and Crispin Sartwell. Blackwell Publishers, 1998. 268–80.

– *Volatile Bodies.* Allen and Unwin, 1994.

Guerrier, Simon. 'Culture Theory: Iain M. Banks's "Culture" as Utopia.' *Foundation* 76 (Summer 1999): 28–38.

Haraway, Donna. 'The Actors Are Cyborg, Nature Is Coyote and the Geography Is Elsewhere: Postscript to "Cyborgs at Large."' Penley and Ross 21–6.

– 'The Biopolitics of Postmodern Bodies: Determinations of Self in Immune System Discourse.' Shildrick and Price 203–14.

– 'A Cyborg Manifesto: Science, Technology, and Socialist-Feminism in the Late Twentieth Century.' *Simians, Cyborgs and Women: The Reinvention of Nature.* Routledge, 1991. 149–81.

– *Modest_Witness@Second_Millennium: FemaleMan©_Meets_OncoMouse™: Feminism and Technoscience.* Routledge, 1997.

– 'The Persistence of Vision.' *Writing on the Body: Female Embodiment and Feminist Theory.* Edited by Katie Conboy, Nadia Medina, and Sarah Stanbury. Columbia University Press, 1997. 283–95.

– *Primate Visions: Gender, Race and Nature in the World of Modern Science.* Routledge, 1989.

– 'The Promises of Monsters: A Regenerative Politics for Inappropriate/d Others.' Grossberg et al. 295–337.

– 'Situated Knowledges: The Science Question in Feminism and the Privilege of Partial Perspective.' *Simians, Cyborgs and Women: The Reinvention of Nature.* Routledge, 1991. 183–201.

Hardesty, William. 'Mercenaries and Special Circumstances: Iain M. Banks's Counter-Narrative of Utopia, *Use of Weapons.*' *Foundation* 76 (1999): 39–47.

Harris, Leslie. 'The Psychodynamic Effects of Virtual Reality.' *Electronic Journal on Virtual Culture* 2.1 (1994). www.monash.edu.au/journals/ejvc/harris.v2n1. 18 February 1999.

Hayles, N. Katherine. 'How Cyberspace Signifies: Taking Immortality Literally.'

*Immortal Engines: Life Extension and Immortality in Science Fiction and Fantasy.*
Edited by George Slusser, Gary Westfahl, and Eric S. Rabkin. University of
Georgia Press, 1996. 111–21.

– *How We Became Posthuman: Virtual Bodies in Cybernetics, Literature, and Information.* University of Chicago Press, 1999.

Hayles, Katherine, David Porush, Brooks Landon, Vivian Sobchack, and J.G.
Ballard. 'In Response to Jean Baudrillard.' *Science Fiction Studies* 18.3 (1991):
321–9.

Haywood, Trevor. 'Global Networks and the Myth of Equality: Trickle Down or
Trickle Away?' Loader, *Cyberspace Divide* 19–34.

Heim, Michael. 'The Erotic Ontology of Cyberspace.' *Cyberspace: First Steps.*
Edited by Michael Benedikt. MIT Press, 1991. 59–80.

– *The Metaphysics of Virtual Reality.* Oxford University Press, 1993.

Hennessy, Rosemary. *Materialist Feminism and the Politics of Discourse.* New York:
Routledge, 1993.

Holden, Rebecca. 'The High Costs of Cyborg Survival: Octavia Butler's *Xenogenesis* Trilogy.' *Foundation* 72 (1998): 49–57.

Holderness, Mike. 'Who Are the World's Information-Poor?' Loader, *Cyberspace
Divide* 35–56.

Hollinger, Veronica. 'Cybernetic Deconstructions: Cyberpunk and Postmodernism.' McCaffery, *Storming the Reality Studio* 203–18.

– 'Feminist Science Fiction: Breaking Up the Subject.' *Extrapolation* 31.3 (1990):
229–39.

– 'Feminist Science Fiction: Construction and Deconstruction.' *Science Fiction
Studies* 16.2 (1989): 223–7.

– 'A New Alliance of Postmodernism and Feminist Speculative Fiction.' *Science
Fiction Studies* 20.2 (1993): 266–72.

– 'The Vampire and the Alien: Variations on the Outsider.' *Science Fiction Studies*
16.2 (1989): 145–60.

– 'Women in SF and Other Hopeful Monsters.' *Science Fiction Studies* 17.2 (July
1990): 129–35.

Hopkins, Patrick D. 'Gender Treachery: Homophobia, Masculinity, and Threatened Identities.' *Race, Class, Gender, and Sexuality: The Big Questions.* Edited by
Naomi Zack, Laurie Shrage, and Crispin Sartwell. Blackwell Publishers, 1998.
168–86.

Hubbard, Ruth, and Elijah Wald. *Exploding the Gene Myth.* Beacon Press,
1993.

Irigary, Luce. *This Sex Which Is Not One.* Cornell University Press, 1985.

James, Edward. *Science Fiction in the Twentieth Century.* Oxford University Press,
1994.

– 'Yellow, Black, Metal and Tentacled: The Race Question in American SF.' *Sci-*

*ence Fiction, Social Conflict and War.* Edited by John Philip Davies. Manchester
University Press, 1990. 26–49.

Jameson, Fredric. 'Cognitive Mapping.' *Marxism and the Interpretation of Culture.*
Edited by Cary Nelson and Lawrence Grosberg. University of Illinois Press,
1988. 347–60.

– *Postmodernism, or, the Cultural Logic of Late Capitalism.* Duke University Press,
1991.

– 'Progress vs. Utopia; or, Can We Imagine the Future?' *Science Fiction Studies* 9.2
(1982): 147–58.

Jenkins, Henry. *Textual Poachers: Television Fans and Participatory Culture.* Routledge, 1992.

Johnson, Mark. 'Embodied Reason.' Weiss and Haber 81–102.

Johnson, Mark, and George Lakoff. *Philosophy in the Flesh.* Basic Books, 1999.

Jones, Gwyneth. 'Aliens in the Fourth Dimension.' *Deconstructing the Starships*
108–19.

– *Deconstructing the Starships: Science, Fiction and Reality.* Liverpool University
Press, 1999.

– 'Fools: The Neuroscience of Cyberspace.' *Deconstructing the Starships* 77–90.

– 'Metempsychosis of the Machine: Science Fiction in the Halls of Karma.'
*Science Fiction Studies* 24 (1997): 1–10.

– *North Wind.* TOR, 1994.

– *Phoenix Café.* TOR, 1998.

– 'Sex: The Brains of Female Hyena Twins.' *Deconstructing the Starships:*
99–107.

– *White Queen.* TOR, 1991.

Kadrey, Richard. 'Reach Out and Touch Someone.' *Shift* (November 1999): 44,
46.

Keller, Evelyn Fox. 'Nature, Nurture and the Human Genome Project.' Kevles
and Hood 281–99.

Kendall, Lori. 'MUDder? I Hardly Know 'Er! Adventures of a Feminist MUD-
der.' *Wired_Women: Gender and New Realities in Cyberspace.* Edited by Lynn Cherney and Elizabeth Reba Weise. Seal Press, 1996. 207–23.

Kevles, Daniel. 'Out of Eugenics: The Historical Politics of the Human
Genome.' Kevles and Hood 3–36.

Kevles, Daniel, and Leroy Hood, eds. *The Code of Codes: Scientific and Social Issues
in the Human Genome Project.* Harvard University Press, 1992.

Kirby, Vicki. *Telling Flesh: The Substance of the Corporeal.* Routledge, 1997.

Krieger, Lisa. 'Nice Mice Just a New Gene Away.' *Edmonton Journal,* 22 August
1999, B2.

Kristeva, Julia. *The Powers of Horror.* Columbia University Press, 1982.

Kroker, Arthur, and Marilouise Kroker, eds. *Digital Delirium.* St Martin's Press, 1997.

– 'Theses on the Disappearing Body in the Hyper-Modern Condition.' *Body Invaders: Panic Sex in America.* Edited by Arthur Kroker and Marilouise Kroker. CultureTexts. St Martin's Press, 1987. 20–34.

Kuhn, Annette, ed. *Alien Zone: Cultural Theory and Contemporary Science Fiction.* Verso, 1990.

Kuhse, Helga, and Peter Singer. *Should the Baby Live? The Problem of Handicapped Infants.* Oxford University Press, 1985.

Kurzweill, Ray. *The Age of Spiritual Machines: When Computers Exceed Human Intelligence.* MIT Press, 1999.

Lacan, Jacques. 'The Mirror Stage.' *Seminar I.* Edited by Jacques-Alain Miler. W.W. Norton, 1988.

Lander, Eric. 'DNA Fingerprinting: Science, Law and the Ultimate Identifier.' Kevles and Hood 191–210.

Latour, Bruno. *We Have Never Been Modern.* Translated by Catherine Porter. Harvard University Press, 1993.

Ledbetter, Mark. *Victims and the Postmodern Narrative or Doing Violence to the Body: An Ethic of Reading and Writing.* St Martin's Press, 1996.

Lefanu, Sarah. *In the Chinks of the World Machine.* Woman's Press 1988.

Le Guin, Ursula. *The Left Hand of Darkness.* Ace Books, 1976 [original text copyright 1969].

Levy, Michael M. 'Ophelia Triumphant: The Survival of Adolescent Girls in Recent SF by Butler and Womack.' *Foundation* 72 (Spring 1998): 34–41.

Loader, Brian. 'Cyberspace Divide: Equality, Agency and Policy in the Information Society.' Loader 3–16.

Loader, Brian, ed. *Cyberspace Divide: Equality, Agency and Policy in the Information Society.* Routledge, 1998.

Lowe, Donald M. *The Body in Late-Capitalist USA.* Duke University Press, 1995.

Luckhurst, Roger. '"Horror and Beauty in Rare Combination": The Miscegenate Fictions of Octavia Butler.' *Women: A Cultural Review* 7.1 (1996): 28–38.

Macpherson, C.B. *The Political Theory of Possessive Individualism: Hobbes to Locke.* Clarendon Press, 1962.

Margulis, Lynn, and Dorion Sagan. *Microcosmos: Four Billion Years of Microbial Evolution.* Touchstone, 1986.

Marx, Karl, and Fredric Engels. *The German Ideology.* International Publications, 1947.

McCaffery, Larry. 'An Interview with William Gibson.' McCaffery, *Storming the Reality Studio* 263–85.

- 'Introduction: The Desert of the Real.' McCaffery, *Storming the Reality Studio* 1–16.

McCaffery, Larry, ed. *Across the Wounded Galaxies: Interviews with Contemporary American Science Fiction Writers.* University of Illinois Press, 1990.

- *Storming the Reality Studio: A Casebook of Cyberpunk and Postmodernism.* Duke University Press, 1992.

McHale, Brian. *Postmodernist Fiction.* Methuen, 1987.

McLuhan, Marshall. *Understanding Media: The Extensions of Man.* McGraw-Hill, 1964.

McRae, Shannon. 'Coming Apart at the Seams: Sex, Text, and the Virtual Body.' *Wired_Women: Gender and New Realities in Cyberspace.* Edited by Lynn Cherney and Elizabeth Reba Weise. Seal Press, 1996. 242–63.

Michaels, Walter Benn. 'Political Science Fiction.' *New Literary History* 31 (2000): 649–64.

Miringoff, Marque-Luisa. *The Social Costs of Genetic Welfare.* Rutgers University Press, 1991.

Miss M. 'An Interview with Pat Cadigan.' *Virtual Futures 96 Datafeed.*1996. http://www.t0.or.at/pcadigan/intervw.htm. Accessed 19 June 2002.

More, Max. 'The Extropian Principles Version 3.0: A Transhumanist Declaration.' 1999. www.extropy.org/extprn3.htm. Accessed 12 November 1999.

Moravec, Hans. *Mind Children: The Future of Robot and Human Intelligence.* Harvard University Press, 1988.

Mouffe, Chantal, and Erneau Laclau. *Hegemony and Socialist Strategy.* Verso, 1985.

Nelkin, Dorothy. 'The Social Power of Genetic Information.' Kevles and Hood 177–90.

Nelkin, Dorothy, and M. Susan Lindee. *The DNA Mystique: The Gene as Cultural Icon.* W.H. Freeman, 1995.

Nelkin, Dorothy, and Laurence Tancredi. *Dangerous Diagnostics: The Social Power of Biological Information.* Basic Books, 1989.

Nietzsche, Friedrich. *Thus Spoke Zarathustra.* Modern Library, 1995.

Nixon, Nicola. 'Cyberpunk: Preparing the Ground for Revolution or Keeping the Boys Satisfied?' *Science Fiction Studies* 19.2 (1992): 219–35.

Osherow, Michelle. 'The Dawn of a New Lilith: Revisionary Mythmaking in Women's Science Fiction.' *NWSA Journal* 12.1 (Spring 2000): 68–83.

Paul, Gregory S., and Earl D. Cox. *Beyond Humanism: CyberEvolution and Future Minds.* Charles River Media, 1996.

Pearson, Jacqueline. 'Where No Man Has Gone Before: Sexual Politics and Women's Science Fiction.' *Science Fiction, Social Conflict and War.* Edited by John Philip Davies. Manchester University Press, 1990. 8–25.

Pearson, Patricia. 'Is That a Mouse in Your Pants ...' *National Post,* 11 October 1999, D1, D2.

Penley, Constance, and Andrew Ross. 'Cyborgs at Large: Interview with Donna Haraway.' Penley and Ross 1–20.

Penley, Constance, and Andrew Ross, eds. *Technoculture.* University of Minnesota Press, 1991.

Penley, Constance, et al., eds. *Close Encounters: Film, Feminism and Science Fiction.* University of Minnesota Press, 1991.

Peppers, Cathy. 'Dialogic Origins and Alien Identities in Butler's *Xenogenesis.*' *Science Fiction Studies* 22 (1995): 47–62.

– '"I've Got You Under My Skin": Cyber(sexed) Bodies in Cyberpunk Fictions.' *Bodily Discursions: Genders, Representations, Technologies.* Edited by Deborah Wilson and Christine Moneera Laennec. State University of New York Press, 1997. 163–85.

Probyn, Elspeth. *Sexing the Self.* Routledge, 1993.

Roberts, Robin. *A New Species: Gender and Science in Science Fiction.* University of Illinois Press, 1993.

– 'Postmodernism and Feminist SF.' *Science Fiction Studies* 17 (1990): 136–51.

Rose, Jacqueline. *Why War? Psychoanalysis, Politics, and the Return to Melanie Klein.* Blackwell Publishers, 1993.

Rose, Nikolas. *Inventing Our Selves: Psychology, Power, and Personhood.* Cambridge University Press, 1996.

Ross, Andrew. 'Cyberpunk in Boystown.' *Strange Weather: Culture, Science, and Technology in an Age of Limits.* Verso, 1991. 137–67.

– 'Hacking Away at the Counterculture.' *Strange Weather: Culture, Science, and Technology in an Age of Limits.* Verso, 1991. 75–100.

Rucker, Rudy. *Freeware.* Avon Books, 1997.

– *Software.* Avon Books, 1981.

– *Wetware.* Avon Books, 1988.

Russ, Joanna. *What Are We Fighting For? Sex, Race, Class and the Future of Feminism.* St Martin's Press, 1998.

Russo, Enzo, and David Cove. *Genetic Engineering: Dreams and Nightmares.* W.H. Freeman, 1995.

Russo, Mary. *The Female Grotesque: Risk, Excess and Modernity.* Routledge, 1995.

Salvaggio, Ruth. 'Octavia Butler and the Black Science Fiction Heroine.' *Black America Literature Forum* 18.2 (1984): 78–81.

Sayers, Janet. 'Adolescent Bodies.' Ussher, *Body Talk* 85–105.

Schatzki, Theodore. 'Practiced Bodies: Subjects, Genders and Minds.' Schatzki and Natter 49–78.

Schatzki, Theodore, and Wolfgang Natter, eds. *The Social and Political Body.* Guilford Press, 1996.

Scholes, Robert. 'The Roots of Science Fiction.' *SF: A Collection of Critical Essays.* Edited by Mark Rose. Prentice-Hall, 1976. 46–56.

Schroeder, Karl. *Permanence.* TOR, 2002.

Shea, Jim. 'Pet Clones Too Doggone Weird.' *Edmonton Journal,* 3 July 1999, F2.

*Shift: For Living in Digital Culture.* November 1999.

Shildrick, Margrit. 'Posthumanism and the Monstrous Body.' *Body and Society* 2.1 (1996):1–16.

Shildrick, Margrit, and Janet Price, eds. *Feminist Theory and the Body: A Reader.* Routledge, 1999.

Silver, Lee M. *Remaking Eden: Cloning and Beyond in a Brave New World.* Avon Books, 1997.

Sinsheimer, Robert. 'An Evolutionary Perspective on Genetic Engineering.' *New Scientist* (20 January 1977): 150.

Slouka, Mark. *War of the Worlds: Cyberspace and the Hi-Tech Assault on Reality.* Basic Books, 1995.

Smith, Paul. *Discerning the Subject.* University of Minnesota Press, 1988.

Smith, Sidonie. *Subjectivity, Identity and the Body: Women's Autobiographical Practices in the Twentieth Century.* Indiana University Press, 1993.

Smith, Stephanie. 'Morphing, Materialism and the Marketing of *Xenogenesis*.' *Genders* 18 (1993): 67–86.

Sobchack, Vivian. 'Beating the Meat/Surviving the Text, or How to Get Out of This Century Alive.' *Cyberspace/Cyberbodies/Cyberpunk.* Edited by Mike Featherstone and Roger Burrows. Sage Press, 1995. 205–14.

Soper, Kate. *Humanism and Anti-Humanism.* HarperCollins, 1986.

Springer, Claudia. 'Muscular Circuitry: The Invincible Armored Cyborg in Cinema.' *Genders* 18 (Winter 1993): 87–101.

Squire, Corrine. 'AIDS Panic.' Ussher, *Body Talk* 50–69.

Star, Susan Leigh. 'From Hestia to Home Page: Feminism and the Concept of Home in Cyberspace.' Braidotti and Lykke 30–46.

Stephenson, Neal. *The Diamond Age.* Bantam Books, 1995.

– *Snow Crash.* Bantam Books, 1992.

Sterling, Bruce. Preface. *Mirrorshades: The Cyberpunk Anthology.* Edited by Bruce Sterling. Arbor House, 1986. vii–xiv.

– *Schismatrix.* Ace Books, 1985.

– 'Unstable Networks.' Kroker and Kroker, *Digital Delirium* 25–37.

Stone, Allucquère Rosanne. *The War of Desire and Technology at the Close of the Mechanical Age.* MIT Press, 1996.

– 'Will the Real Body Please Stand Up? Boundary Stories about Virtual Cultures.' *Cyberspace: First Steps.* Edited by Michael Benedikt. MIT Press, 1991. 81–118.

Stoppard, Janet. 'Women's Bodies, Women's Lives and Depression.' Ussher, *Body Talk* 10–32.

Suvin, Darko. 'On Gibson and Cyberpunk SF.' *Foundation* 46 (1989): 40–51.

Terry, Jennifer. 'The Seductive Power of Science in the Making of Deviant Subjectivity.' *Posthuman Bodies.* Edited by Ira Livingston and Judith Halberstrom. Indiana University Press, 1995. 135–61.

Turkle, Sherry. *Life on the Screen: Identity in the Age of the Internet.* Touchstone, 1995.

Ullman, Ellen. *Close to the Machine: Technophilia and Its Discontents.* City Lights Books, 1997.

Ussher, Jane. 'Framing the Sexual "Other".' Ussher, *Body Talk* 131–58.

Ussher, Jane, ed. *Body Talk: The Material and Discursive Regulation of Sexuality, Madness and Reproduction.* Routledge, 1997.

Vinge, Vernor. 'What Is the Singularity.' 1993. *VISION-21 NASA Symposium.* http://www.ugcs.caltech.edu/~phoenix/vinge/vinge-sing.html. 19 September 2001.

Watson, James. 'A Personal View of the Project.' Kevles and Hood 164–73.

Weinstone, Ann. *Avatar Bodies: A Tantra for Posthumanism.* University of Minnesota Press, 2003.

Weiss, Gail. 'The Abject Borders of the Body Image.' Weiss and Haber 41–60.

– *Body Images: Embodiment as Intercorporeality.* Routledge, 1999.

Weiss, Gail, and Honi Fern Haber, eds. *Perspectives on Embodiment: The Intersections of Nature and Culture.* Routledge, 1999.

Wexler, Nancy. 'Clairvoyance and Caution: Repercussions from the Human Genome Project.' Kevles and Hood 211–43.

White, Eric. 'The Erotics of Becoming: *Xenogenesis* and *The Thing*.' *Science Fiction Studies* 20 (1993): 394–408.

Whalen, Terence. 'The Future of a Commodity: Notes toward a Critique of Cyberpunk and the Information Age.' *Science Fiction Studies* 19.1 (1992): 75–88.

Williams, Raymond. *Problems in Materialism and Culture: Selected Essays.* Verso, 1980.

Wolmark, Jenny. *Aliens and Others: Science Fiction, Feminism, and Postmodernism.* University of Iowa Press, 1994.

Womack, Jack. *Random Acts of Senseless Violence.* Grove Press, 1993.

X, Robert Adrian. 1997. 'Infobahm Blues.' Kroker and Kroker, *Digital Delirium* 84–8.

Yanarella, Ernest J., and Herbert G. Reid. 'From "Trained Gorilla" to "Human-
    ware": Repoliticizing the Body-Machine Complex between Fordism and Post-
    Fordism.' Schatzki and Natter 181–220.
Young, Iris. *Throwing Like a Girl and Other Essays.* Indiana University Press, 1990.
Zaki, Hoda. 'Utopia, Dystopia, and Ideology in the Science Fiction of Octavia
    Butler. *Science Fiction Studies* 17 (1990): 239–51.

# Index

Adam, Alison, 108
agency, 4–6, 157–9, 215n4; and the
  body, 19, 69–70; and cyberpunk,
  103, 111; and identity, 14–16, 29,
  58; and interpellation, 31, 37, 75–6,
  139–43, 169–70; and liberal
  humanism, 12–14; and materiality,
  166; and subjectivity, 22, 70; and
  technology, 118–22; through reading and writing, 24–5, 138–54,
  158–61, 164–70
AIDS. *See* body
Alaimo, Stacy, 58
Aldiss, Brian, 195n19
Aldridge, Susan, 197n9
Althusser, Louis, 20, 22, 29–31. *See*
  *also* interpellation
animal, 39, 64, 91, 125–8, 186,
  188
Appleyard, Bryan, 197n10
artificial intelligence, 79, 91–2,
  113–14, 115, 171–2, 203n1
Attebury, Brian, 20
Atzori, Paulo, 118, 176
autobiography. *See* agency, through
  reading and writing
autonomy. *See* agency

Balsamo, Anne, 105, 115–17, 120–1,
  181–2, 217n3
Banks, Iain M., 23, 79–101, 181,
  203n3, 203n5; *Consider Phlebas*, 23,
  80, 90–2, 94–5, 203n7, 204n8; *Exces-*
  *sion*, 99; *Look to Windward*, 99–100;
  *The Player of Games*, 23, 80, 92, 93,
  95–9, 100, 204n9, 205n10, 205n11;
  'Some Notes on the Culture,' 80–1,
  88–91, 93–4, 203n1; 'The State of
  the Art,' 80, 89; *Use of Weapons*, 23,
  80, 86–7, 92, 97, 203n4, 206n13
Barglow, Raymond, 109–10, 119–21
Barnard, Jeff, 197n7
Barr, Marlene, 20, 71
Bear, Greg, 54
Belkin, Lisa, 197n8, 198n14
Best, Stephen, 60, 219n17
biopower. *See* eugenics
birth, 199n17
body, the (*see also* embodiment;
  eugenics)
– commodified, 108
– and community, 183–4
– and difference, 58, 70, 92, 93, 149–
  50, 184–6, 201n24; class, 8–9, 144–
  8, 205n10; gender, 17, 23, 46–7, 50,

www.ingramcontent.com/pod-product-compliance
Lightning Source LLC
Chambersburg PA
CBHW030241030426
42336CB00009B/193